Metabolic Aspects
of Lipid Nutrition
in Insects

Also of Interest

The Biology of Social Insects: Proceedings of the Ninth Congress of the International Union for the Study of Social Insects, **edited by Michael D. Breed, Charles D. Michener, and Howard E. Evans**

Orthopteran Mating Systems: Sexual Competition in a Diverse Group of Insects, **edited by Darryl T. Gwynne and Glenn K. Morris**

Insect Behavior: A Sourcebook of Laboratory and Field Exercises, **edited by Janice R. Matthews and Robert W. Matthews**

Pest Control: Cultural and Environmental Aspects, **edited by David Pimentel and John H. Perkins**

World Food, Pest Losses, and the Environment, **edited by David Pimentel**

A Westview Science Study

Metabolic Aspects
of Lipid Nutrition in Insects
edited by T. E. Mittler and R. H. Dadd

Our understanding of the physiological function of insect essential lipids has long been flawed by major uncertainties. It was discovered long ago that dietary sterol is a necessary nutrient for all insects, which radically sets them apart from the vertebrates in terms of qualitative nutrient requirements. Because of the physiological importance of sterol as a molting hormone precursor in insects and the implications of this for the development of new insecticides, a wealth of investigation into insect sterol metabolism followed, covering both the ways in which insects convert diverse food-plant sterols into the major tissue sterols and how these in turn are metabolized into the ecdysone molting hormones.

However, for the classes of essential lipid nutrients required by vertebrates, research dealing with insects has been scant and, more often than not, rather indeterminate. Many, but by no means all, insects studied appear to require essential fatty acids, though virtually nothing has been found out about the metabolism or essential physiological function of these acids. Excepting vitamin A, needed for insect vision, the various vertebrate fat-soluble vitamins appear to have no significance for insect physiology, and results of the occasional attempts to demonstrate functions for them in growth and development have in most cases been tantalizingly equivocal.

In recent years some notable advances were made in the study of essential fatty acids and fat-soluble vitamins in insects, and work on insect sterol nutrition and metabolism continues with ever-increasing sophistication. The contributors to this book summarize, discuss, and speculate on these issues. Their work is based on papers presented at the 1980 World Congress of Entomology at Kyoto, Japan.

Dr. Mittler and Dr. Dadd are entomologists at the University of California, Berkeley.

Metabolic Aspects of Lipid Nutrition in Insects

edited by T. E. Mittler and R. H. Dadd

Routledge
Taylor & Francis Group

LONDON AND NEW YORK

First published 1983 by Westview Press

Published 2018 by Routledge
52 Vanderbilt Avenue, New York, NY 10017
2 Park Square, Milton Park, Abingdon, Oxon OX14 4RN

Routledge is an imprint of the Taylor & Francis Group, an informa business

Library of Congress Cataloging in Publication Data
Main entry under title:
Metabolic aspects of lipid nutrition in insects.
  (A Westview science study)
  Includes index.
  1. Insects--Physiology. 2. Lipids--Metabolism.
I. Mittler, Thomas E. II. Dadd, R. H. III. Title: Lipid nutrition in insects.
IV. Series.
QL495.M47   1983   595.7'0133   82-21892

ISBN 13: 978-0-367-01912-9 (hbk)
ISBN 13: 978-0-367-16899-5 (pbk)

Contents

viii

Preface

Physiological understanding of the essential lipid nutrients required by insects is flawed by major uncertainties except with regard to the universal sterol requirement. It was realized long ago that dietary sterol is necessary for all insects, a situation that radically sets them, and probably arthropods in general, apart from the vertebrates in terms of qualitative nutrient requirements. In truth, without the discovery in the mid-1930s of the need for dietary cholesterol by blowfly larvae, subsequent development of the discipline of insect nutrition would have been severely delayed, since it would have been difficult or impossible to achieve good growth on synthetic diets with any insect except those containing steroidogenic symbiotes or those whose synthetic diets carried cryptic sterol impurities in other dietary ingredients. Because of the physiological importance of sterol as a necessary precursor for moulting hormones in insects and the implications of this for the attempt to develop novel insecticides, a wealth of investigation into insect sterol metabolism followed. This work examined the various ways in which insects convert diverse food-plant sterols into the major insect tissue sterols, in most cases predominantly cholesterol, and how this latter in turn is metabolized into the ecdysone moulting hormones.

By contrast, research on insects that deals with the classes of essential lipid nutrients required by vertebrates has been scant and, more often than not, rather indeterminate in outcome. For many years after discovery of the insect sterol requirement it was thought that no other lipid nutrients were needed, though it was understood that fats and oils, if present in food, could be utilized as contributors to overall energetic requirements. During the mid-1940s it was shown that some insects require essential polyunsaturated fatty acids. Probably most, though by no means all, are now considered to have this requirement. However, until recently, virtually nothing was known of the metabolism or physiological functions of essential fatty acids in insects. Excepting vitamin A, which is now well understood to be dietarily necessary for normal insect vision, the various vertebrate fat-soluble vitamins generally appeared of dubious or no significance in insect physiology, and results of the occasional attempts to demonstrate functions for them in growth and development were in most cases negative or tantalizingly equivocal.

Recent years have seen some notable advances in the study of essential fatty acids and fat-soluble vitamins needed by insects, while work on insect sterol nutrition and metabolism continues with ever-increasing sophistication. These topics formed the core of a symposium convened at the 1980 International Congress of Entomology at Kyoto, Japan, to summarize, discuss, and speculate on these issues. This book is compiled from papers presented at the Kyoto symposium, updated in many cases during the following year, and augmented by material from some authors who, though unable to attend the meeting, had valuable contributions that reduce the lacunae inevitable in a multi-author treatment of a wide-ranging topic such as this. We believe that after a rather dormant period, sterol studies excepted, the issue of vitamin-like lipid requirements in insect nutrition is currently showing signs of revivification. Our hope is that despite some omissions of coverage -- retrospectively one regrets that vitamin A slipped through our net -- this volume will serve to crystallize current thinking in the area for those interested in furthering its future development.

T.E.M.
R.H.D.

1
Comparative Sterol Metabolism in Insects

*J. A. Svoboda and M. J. Thompson**

ABSTRACT

Most insects have a critical need for dietary sterol and the utilization and metabolism of sterols varies considerably from species to species. Adaptations to numerous ecological niches and widely different food sources utilized by insects apparently have been accompanied by corresponding physiological and biochemical adaptation in sterol utilization and metabolism. Some of the more interesting differences between species in this area of insect biochemistry are discussed.

INTRODUCTION

Investigations of normal sterol metabolism with a number of insect species that occupy a variety of ecological niches have revealed differences between species that provide evidence that there is considerably more diversity in the area of biochemistry of sterols in the Class Insecta than was previously thought to exist. Research into the normal utilization and metabolism of dietary sterols in insects has revealed that it is very difficult to make generalizations concerning this important area of insect biochemistry (Svoboda et al., 1978). The fate of dietary sterols has been of interest to insect biochemists since it was discovered that insects require a dietary source of sterol (Hobson, 1935), that they are incapable of de novo biosynthesis of the steroid nucleus (Bloch et al., 1956), and that certain insects are able to dealkylate the C-24 position of the sterol side chain (Clark and Bloch, 1959; Robbins et al., 1962; Robbins et al., 1971).

* Insect Physiology Laboratory Agricultural Research Service, U.S. Department of Agriculture, Beltsville, MD 20705.

At least one essential physiological role for cholesterol was demonstrated when cholesterol was first shown to be an ecdysteroid (molting hormone) precursor (Karlson and Hoffmeister, 1963). The report of the conversion of ^{3}H-sitosterol to ^{3}H-cholesterol in the German cockroach Blattella germanica (Robbins et al., 1962) and, subsequently, by several other species of omnivorous and phytophagous insects (Robbins et al., 1971) provided impetus for considerable research on the utilization and metabolism of dietary C_{28} and C_{29} phytosterols. A number of intermediates involved in the dealkylation and conversion of phytosterols to cholesterol, particularly in the tobacco hornworm (Manduca sexta), were subsequently isolated and identified, and considerable progress has been made toward elucidating these pathways in several omnivorous and phytophagous species (Svoboda et al., 1978). However, as insects that undergo various types of development and occupy a diversity of ecological niches were examined, more and more differences between species in the area of sterol metabolism were discovered.

For a number of years it was held that insects could be neatly divided into two general categories with respect to plant sterol metabolism: with the zoophagous species, generally held to be unable to dealkylate and convert plant sterols to cholesterol in one group and the phytophagous and omnivorous species that can dealkylate substituents at C-24 of the sterol side chain in a second group. Over the past several years, studies in our laboratory have revealed some interesting differences that exist between certain species of phytophagous insects. A number of these unexpected variations in utilization and metabolism of dietary sterols in several species of insects suggest provoking implications of phylogenetic relationships between certain species and will be discussed in this paper.

STEROL METABOLISM IN THE TOBACCO HORNWORM

Although we had known for a number of years that sitosterol is converted to cholesterol in several omnivorous and phytophagous species of insects, nothing was known of the intermediate steps in this process. The first intermediate involved in the conversion of phytosterols to cholesterol in an insect was isolated and identified from tobacco hornworms that had been fed sitosterol. ^{3}H-Desmosterol was isolated from the sterols of ^{3}H-sitosterol-fed insects (Svoboda et al., 1967) and later found to be a constant metabolite of sitosterol and several other C_{28} and C_{29} phytosterols, including campesterol, stigmasterol, brassicasterol, 22,23-dihydrobrassicasterol, 24-methylenecholesterol, and fucosterol (Svoboda et al., 1975), all of which are converted to cholesterol by the tobacco hornworm. Desmosterol is also an intermediate in the conversion of sitosterol to cholesterol in other phytophagous and omnivorous species, as shown by comparative studies with the firebrat (Thermobia domestica), the German cockroach, the American cockroach (Periplaneta americana), the corn earworm (Heliothis zea), and the fall armyworm (Spodoptera frugiperda) (Svoboda and Robbins, 1971).

We have investigated the metabolism of the three most common phytosterols, sitosterol, campesterol, and stigmasterol, in the tobacco hornworm in considerable depth. Another intermediate found to be involved in the conversion of sitosterol to cholesterol is fucosterol, which has a 24-ethylidine group (Svoboda et al., 1975). An analogous pathway was demonstrated for the conversion of campesterol to cholesterol in that 24-methylenecholesterol precedes desmosterol in the pathway to cholesterol (Svoboda et al., 1975). We also isolated and identified 5,22,24-cholestatrienol as a normal intermediate in the conversion of stigmasterol to cholesterol in the hornworm. This was the first isolation of this sterol from a biological source (Svoboda et al., 1975), and it was determined that in its conversion to cholesterol the Δ^{22}-bond is reduced first and that the Δ^{24}-bond is essential for substrate specificity.

This work, done primarily with the tobacco hornworm, was the foundation for subsequent studies with other species. The metabolic pathways outlined in Figure 1, however, did not turn out to be a blueprint for all insects generally categorized as phytophagous because they feed on plants or plant material.

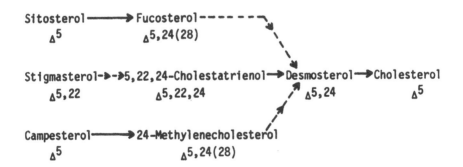

FIGURE 1. Pathways of dealkylation and conversion of C_{28} and C_{29} phytosterols to cholesterol in the tobacco hornworm.

STEROL METABOLISM IN THE CONFUSED FLOUR BEETLE

A previous report that 7-dehydrocholesterol constituted over half of the total tissue sterols of the confused flour beetle (<u>Tribolium</u> <u>confusum</u>) (Beck and Kapadia, 1957) prompted us to examine in depth the normal
metabolism of several radiolabeled phytosterols in this
insect. All omnivorous or phytophagous insects that we
had examined prior to that time had primarily Δ^5-sterols
in their tissues, and 7-dehydrocholesterol was normally
only a minor (1% of total sterols) component of the total sterols. Ecdysteroids have a Δ^7-bond, and 7-dehy-
drocholesterol has been reported to be converted to 20-
hydroxyecdysone in <u>Calliphora</u> <u>stygia</u> (Galbraith et al.,
1970). No other role is known for a Δ^7-bond, but we
did find that 7-dehydrocholesterol constituted as much
as 50% of the sterols of the confused flour beetle when
radiolabeled sitosterol, campesterol, stigmasterol or

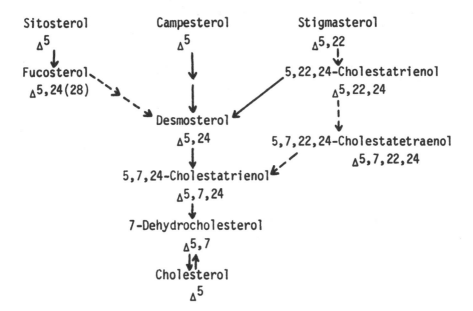

FIGURE 2. Pathways of dealkylation and conversion of
C_{28} and C_{29} phytosterols to cholesterol in the confused flour beetle.

desmosterol were fed (Svoboda et al., 1972). The only other major sterol produced from each of these dietary sterols was cholesterol, and studies in which labeled cholesterol or 7-dehydrocholesterol acetate were fed verified that an equilibrium between cholesterol and 7-dehydrocholesterol is maintained in the tissues of this insect.

Desmosterol, fucosterol, and 5,22,24-cholestatrienol were similarly involved in the pathways of conversion of the phytosterols to cholesterol in the confused flour beetle as in the previously elucidated pathways in the tobacco hornworm (Figure 2). Also, a new sterol intermediate, 5,7,24-cholestatrienol was shown to be a normal intermediate in the conversion of Δ^5- or $\Delta^{5,22}$-phytosterols to cholesterol in this insect, representing an important variation from what had been found in our previous studies on sterol metabolism. T. confusum also produced a related compound, 5,7,22,24-cholestatetraenol, from dietary stigmasterol, when an azasteroid inhibitor was fed simultaneously (Svoboda et al., 1972).

The purpose of the high levels of 7-dehydrocholesterol in T. confusum remains obscure, since the only known specific role for this sterol is as a precursor of the ecdysteroids and only a small fraction of that found in this insect would be sufficient for this purpose. We have also found that 15-20% of the sterols of another flour beetle, Tenebrio molitor, consists of 7-dehydrocholesterol; thus some adaptation to their ecological niches may be responsible for the unusual sterol composition in these flour beetles.

STEROL METABOLISM IN THE KHAPRA BEETLE

Another stored-product insect, the khapra beetle, Trogoderma granarium, is able to utilize dietary sterols in an unexpected manner. When this insect was fed a diet consisting only of wheat and brewer's yeast, it did not dealkylate the dietary C_{28} or C_{29} phytosterols to form C_{27} sterols such as cholesterol (Svoboda et al., 1979). Instead, there seemed to be some selective uptake of cholesterol and campesterol, resulting in an enrichment of these sterols in the insect tissues as compared to the dietary sterols. These results bring to mind the selective uptake and/or retention of sterols previously demonstrated with the house fly, Musca domestica (Thompson et al., 1963).

No measurable amount of cholesterol was found in mature khapra beetle larvae that had been fed radiolabeled sitosterol, stigmasterol or desmosterol (Svoboda et al., 1980). Similar results were obtained when each of these labeled sterols was injected into last-instar larvae. Thus, it was concluded that T. granarium does

not dealkylate and convert C_{29} phytosterols to cholesterol. Also of considerable interest is the fact that,
unlike Tribolium or T. molitor, which have considerably
higher levels of 7-dehydrocholesterol than most insects,
no 7-dehydrocholesterol was detected in T. granarium.

The fate of plant sterol in the khapra beetle differs
markedly from what was thought to be the normal fate of
plant sterols in a phytophagous insect. T. granarium is
very different in this respect from T. confusum, although
the two species occupy fairly similar ecological niches.
With respect to sterol utilization, the khapra beetle is
quite comparable to another dermestid, Dermestes maculatus, whose sterol metabolism was examined extensively
(Clark and Bloch, 1959a). These results dramatically
emphasize the pitfalls of making broad generalizations
concerning sterol metabolism in insects.

STEROL METABOLISM IN THE MEXICAN BEAN BEETLE

Some highly unusual results were obtained in studies
on sterol metabolism in the Mexican bean beetle, Epilach-
na varivestis (Svoboda et al., 1974). This species is
interesting from a phylogenetic standpoint since it is a
member of the family Coccinellidae which includes the
ladybug beetles. Most of the members of this family are
predacious, and, therefore, beneficial insects; however,
the subfamily Epilachninae, to which the Mexican bean
beetle belongs, is composed of phytophagous species, some
of which are important pests of agriculture.

The sterols from prepupae and adults reared on soybean leaves consisted largely of saturated sterols, including cholestanol, campestanol, and stigmastanol. The
prepupal and adult sterols contained 72 and 77% stanols,
respectively, but 5% of the sterols in either stage was
cholesterol. Soybean sterols, on the other hand, consist
largely of campesterol, stigmasterol and sitosterol (>98%
of total) and contain 1% cholesterol and only ca. 1.4%
total saturated sterols. Significantly, lathosterol
(Δ^7-cholestenol) was found to be a major sterol of the
Mexican bean beetle, comprising 16.0 and 11.8% of the
total sterols of prepupae and adults, respectively.

This is the first example of a host-plant-reared insect found to have significant levels of saturated sterols. Usually in plant-feeding insects, cholesterol constitutes 80-90% of the tissue sterols, so bean beetles
do not convert phytosterols to cholesterol to the extent
expected. It is also the first report of lathosterol
being identified as one of the major sterol components
of a plant-fed insect.

The foregoing results, plus data from feeding studies
with radiolabeled cholesterol, cholestanol, stigmasterol,
stigmastanol and sitosterol coated on soybean leaves,

allowed us to propose the scheme in Figure 3 for the metabolism of plant sterols in the Mexican bean beetle (Svoboda et al., 1975a). The Δ^5-dietary phytosterols are first reduced to stanols, which are then dealkylated to produce cholestanol; the Δ^7-bond is then introduced to produce lathosterol. The dietary stanols are not absorbed, dealkylated, and converted to lathosterol as readily as dietary Δ^5-sterols; thus, the 5,6 double bond is necessary for good utilization by the bean beetle.

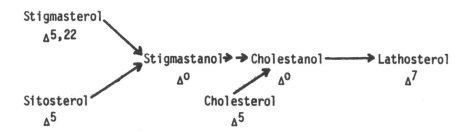

FIGURE 3. Utilization of dietary phytosterols in the Mexican bean beetle.

COMPARISON OF STEROL METABOLISM IN MEXICAN BEAN BEETLE AND COCCINELLA SEPTEMPUNCTATA

Comparing related insects with different feeding habits, such as phytophagous and predacious species of Coccinellidae, has provided even more interesting information on differences in sterol utilization and metabolism between species. The sterols of the predacious coccinellid, Coccinella septempunctata, were compared with those of the phytophagous Mexican bean beetle to determine whether the unique pattern of sterol utilization and metabolism of the bean beetle is characteristic of coccinellids in general or is peculiar to phytophagous species (Svoboda and Robbins, 1979).

The sterols of C. septempunctata were largely those that would be expected from a predacious insect fed a normal diet of phytophagous insects such as aphids (Table 1). Nearly half the total sterols consisted of cholesterol; and campesterol, stigmasterol, and sitosterol occurred in lesser concentrations. Only about 2.7% of the sterols of C. septempunctata were saturated sterols; and lathosterol was not detected in this species, whereas it was present in significant concentrations in the bean beetle. On the other hand, 7-dehydrocholesterol was present in the predatory species (3.7%) but not in the phytophagous species. The sterols of C. septempunctata could represent the normal sterol pattern for predacious coccinellids if phytophagy arose secondarily in a species such as the Mexican bean beetle, whose ancestors may have originally been predacious. These and other studies with related insects having differing feeding habits may provide insight into an important area of biochemical adaptation in relation to phylogeny and speciation within related groups of insects.

TABLE 1. Relative percentages of total saturated and unsaturated sterols from adults of the Mexican bean beetle and Coccinella septempunctata.

Sterol	Mexican bean beetle	Coccinella septempunctata
Saturated Sterols		
Cholestanol	50.7	1.3
Campestanol	6.0	0.6
Stigmastanol	20.3	0.8
Total	77.0	2.7
Unsaturated Sterols		
Cholesterol	4.5	46.4
Δ^7-Cholestenol	11.8	-
7-Dehydrocholesterol	-	3.7
Campesterol	T*	12.2
Δ^7-Campestenol	2.0	-
Stigmasterol	1.4	5.5
Sitosterol	2.3	29.5
Δ^7-Stigmastenol	1.0	-
Total	23.0	97.3

*Trace detectable

STEROL METABOLISM IN THE MILKWEED BUG

An equally unexpected mechanism of plant sterol utilization in a phytophagous insect was discovered in studies with the large milkweed bug, Oncopeltus fasciatus
(Svoboda et al., 1977). The sterols from 4-day-old eggs
and dietary sunflower seeds, as well as those from adult
male and female milkweed bugs reared on sunflower seeds
were compared. Only a small amount (<1%) of cholesterol
was identified in the insect sterols, and this amount
could be accumulated by selective uptake, as discussed in
the earlier section on the khapra beetle. The dietary
sterols underwent little alteration in the milkweed bug
either before or after being incorporated into the insect
tissues or when subsequently transferred to the eggs.
Thus, O. fasciatus is another phytophagous species that
does not dealkylate C_{28} and C_{29} phytosterols to any
measurable extent. Support for this premise was obtained
when radiolabeled campesterol or sitosterol was injected
into milkweed bug nymphs and no labeled cholesterol was
produced.

Of considerable interest is the fact that makisterone
A, a C_{28} ecdysteroid with a 24-methyl group, is the
major ecdysteroid of milkweed bug eggs (Kaplanis et al.,
1975). To date, the evidence on hand from our sterol
metabolism studies indicates that the C_{28} sterol campesterol could well be a precursor of makisterone A in
this species.

STEROL METABOLISM IN THE HONEY BEE

It has long been known that 24-methylenecholesterol
is the major sterol of the honey bee, Apis mellifera
(Barbier and Schindler, 1959). 24-Methylenecholesterol
is commonly found as a major sterol of pollen (Barbier
et al., 1960; Hugel, 1962; Standifer et al., 1968), but
it was not known whether that present in honey bee tissues was derived in part or entirely from dietary pollen
sources or from dealkylation and conversion of phytosterols. We undertook an in-depth study of the utilization
and metabolism of a number of dietary sterols, including
cholesterol, campesterol, sitosterol, stigmasterol, and
24-methylenecholesterol (Herbert et al., 1980). Of the
various sterol-supplemented diets tested, those with
added cholesterol or 24-methylenecholesterol best supported the production of sealed brood and survival of worker
bees.

The sterols of prepupae reared by workers fed chemically defined synthetic diets, each containing one of the
above sterols or no sterol supplement, were examined
(Svoboda et al, 1980a). In every case, 24-methylenecho-
lesterol was the major prepupal sterol, and significant

levels of sitosterol and isofucosterol were also present
in each sample. There was a significant increase in the
relative percentage of the respective dietary sterol in
prepupal sterols, but the levels of cholesterol were very
low (0.4-2.2% of total sterol) except in the cholesterol
supplemented diet (16.1-17.2%). In addition, small quantities of desmosterol were identified in each prepupal
sample. These findings suggested some unusual aspects
of sterol utilization and metabolism in yet another phytophagous insect. Obviously, there was little, if any,
evidence for conversion of C_{28} or C_{29} phytosterols
to cholesterol by the larvae. It was also clear that a
major portion of the sterols made available to the larvae
was derived from the sterol pools of the workers rather
than from the diet that the workers were fed. In addition to supplying the larvae with large amounts of 24-
methylenecholesterol, the workers also provide smaller,
but fairly constant amounts of both sitosterol and isofucosterol. Sterols from newly emerged workers from hives
in which the bees foraged in the wild had levels of cholesterol and desmosterol similar to what we found in the
prepupae from the test diets. Apparently, therefore,
cholesterol and desmosterol are also passed along to the
brood from the workers.
Results from subsequent studies with radiolabeled
campesterol, sitosterol, and 24-methylenecholesterol fed
in chemically defined diets to honey bee workers verified
these results and also indicated labeled campesterol or
sitosterol was not converted to 24-methylenecholesterol
or cholesterol (Svoboda et al., 1981). Further, the
queen also provides sterols from her endogenous pools to
the brood and these sterols are replenished from the
worker diet.
So the honey bee possesses another quite unique mechanism for sterol utilization. The worker and the queen
both have a great propensity for making significant quantities of certain sterols, namely 24-methylenecholester-
ol, sitosterol, and isofucosterol, available to the growing larvae.

STEROL METABOLISM IN THE YELLOWFEVER MOSQUITO

It has been held for some time that dipterous insects
are unable to dealkylate the C-24 position of the phytosterol side chain. The only definitive studies that had
been done on Diptera were with larvae and adults of the
house fly Musca domestica (Kaplanis et al., 1963; Kaplanis et al., 1965). Recently, we examined the fate of [14]C-
sitosterol fed to yellowfever mosquito (Aedes aegypti)
larvae by coating the sterol on the dry diet components
of a modified mosquito diet (Svoboda et al., unpublished). Well over 40% of the recovered radioactivity

from pupal sterols was associated with cholesterol, but about 60% of the mass of the pupal sterols was identified as cholesterol. The females of this species require a blood meal for egg production, and, in this way, they ingest considerable cholesterol. Considering that their larval feeding habits include ingestion of plant material, it is likely that both male and female yellowfever mosquitoes are capable of dealkylating and converting sitosterol to cholesterol.

This is the first demonstration of a dipterous species converting a phytosterol to cholesterol. It will be necessary to resolve the discrepancy in specific activity of recovered cholesterol through aseptic studies in order to isolate the source of any de novo biosynthesis of cholesterol, since the very low levels of cholesterol in the diet components could not account for the dilution that occurs. We have definitely shown, however, that A. aegypti can convert sitosterol to cholesterol, and further in-depth studies of sterol metabolism in this insect should reveal considerably more interesting information.

SUMMARY

Our research on sterol metabolism in insects has revealed considerable diversification and several interesting biochemical variations in the means by which these insects satisfy their sterol requirement.

We have substantiated that some phytophagous insects readily dealkylate plant sterols to form C_{27} sterols and that others are able to directly utilize dietary cholesterol as a precursor to form the much more structurally complex ecdysteroids. To date, no lepidopterous insects examined in detail have been found to deviate from what was considered to be the normal dealkylation and conversion of phytosterols to cholesterol in insects such as the tobacco hornworm (Svoboda et al., 1978). Omnivorous species such as the German cockroach, the American cockroach, and the firebrat (Svoboda and Robbins, 1971) also have the ability to make these conversions by way of the same metabolic pathways, although not as efficiently as the lepidopterous species.

We found an interesting variation on this theme in studies with the confused flour beetle. T. confusum larvae generate very high levels of 7-dehydrocholesterol (up to 50% of the total tissue sterols; Svoboda et al., 1972), whereas this sterol usually constitutes a very small percentage of the total sterols in other insects. Certainly, the fact that another stored-product insect, the khapra beetle, is unable to convert plant sterols to cholesterol or 7-dehydrocholesterol was an unexpected discovery (Svoboda et al., 1979; Svoboda et al., 1980).

C_{28} and C_{29} Phytosterols $\longrightarrow$ Cholesterol
Tobacco hornworm, other Lepidoptera, and certain omnivorous species
Yellow fever mosquito

C_{28} and C_{29} Phytosterols $\longrightarrow$ 7-D $\rightleftharpoons$ Cholesterol
Confused flour beetle

C_{28} and C_{29} Phytosterols $\longrightarrow$ Cholestanol + Lathosterol
Mexican Bean Beetle (Family Coccinellidae)

Utilizes sterols of prey
Coccinella septempunctata (Family Coccinellidae)

C_{28} and C_{29} Phytosterols $\nrightarrow$ Cholesterol
Khapra beetle
Milkweed bug
Honey bee

FIGURE 4. Summary of dietary phytosterol utilization in insects discussed in this paper.

The fact that a leaf-feeding insect such as the Mexican bean beetle does not produce cholesterol from phytosterols, but rather primarily cholestanol plus other C_{28} and C_{29} stanols (Svoboda et al., 1974; Svoboda et al., 1975a) was extremely interesting for several reasons. First of all, it shattered the notion that all leaf-feeding insects might utilize dietary sterols similarly. It also revealed a new metabolic scheme for the utilization of phytosterols in insects that is unique to this species among all the insect species that have been studied in this respect to date. This information also enabled us to compare sterol utilization in two members of the same family with very different feeding habits-- the Mexican bean beetle and the predacious Coccinella septempunctata (Svoboda and Robbins 1979), and to show significant differences between them.

The studies with the milkweed bug revealed another unexpected facet of sterol utilization in insects. This species also feeds on plant material but neither produces cholesterol from C_{28} and C_{29} phytosterols nor dealkylates the C-24 alkyl groups (Svoboda et al., 1977). It is closest in this regard to the khapra beetle among the insects we have studied. Interestingly, it also is essentially a seed feeder.

The honey bee belongs to the general classification of phytophagous insects, but has adapted to utilizing dietary sterols and selectively transferring dietary and endogenous sterols to the brood in the hive (Svoboda et al., 1981; Svoboda et al., 1980a). Certainly this must be the most efficient means of providing dietary sterols to the brood. It will be of interest to conclusively determine what molting hormones are present in the honey bee to see whether the honey bee may have adapted to utilizing a C_{28} or C_{29} phytosterol as an ecdysteroid precursor, just as the milkweed bug apparently has done in utilizing campesterol as a precursor of makisterone A.

Finding that the yellowfever mosquito larva dealkylates and converts phytosterols to cholesterol provides yet another ramification to the story of sterol metabolism in insects. This finding indicates that certain dipterous insects are able to utilize plant sterols in this way even though previous thorough studies with the housefly had shown it to be unable to make this conversion. However, previous reports on the sterol content of Drosophila pachea would indicate that it may be capable of dealkylation (Goodnight and Kircher, 1971; Heed and Kircher, 1965).

Although we have obtained considerable information on the utilization and metabolism of dietary phytosterols, and although we have made progress in our understanding of neutral sterol metabolism, steroid metabolism in insects remains a fruitful area of investigation in insect biochemistry. Certainly, the differences between insect species that we have observed with respect to sterol metabolism, as summarized here, relate to phylogeny or adaptation and indicate that research with insects will continue to be an informative area of comparative steroid biochemistry. It must be stressed that insects, because they lack the ability to biosynthesize the steroid nucleus, do provide us with a nearly ideal experimental system for studying utilization and metabolism and the effect of steroid structure on function in a vast number of species occupying an unparalleled variety of habitats.

REFERENCES

Barbier, M., M.F. Hugel, and E. Lederer. (1960) Isolation of 24-methylenecholesterol from the pollen of different plants. Soc. Chim. Bull. 42: 91-97.

Barbier, M. and O. Schindler. (1959) Isolierung von 24-methylen cholesterin aus Königinnen und Arbeiterinnen der Honigbiene (Apis mellifica L.). Helv. Chim. Acta 42: 1998-2005.

Beck, S.D. and G.G. Kapadia. (1957) Insect nutrition and metabolism of sterols. Science 126: 258-59.

14

Bloch, K., R.G. Langdon, A.J. Clark, and G. Fraenkel. (1956) Impaired steroid biogenesis in insect larvae. Biochim. Biophys. Acta 21: 176.

Clark, A.J. and K. Bloch. (1959) Conversion of ergosterol to 22-dehydrocholesterol in Blattella germanica. J. Biol. Chem. 234: 2589-93.

Clark, A.J. and K. Bloch. 1959a. Function of sterols in Dermestes vulpinus. J. Biol. Chem. 234: 2583-88.

Galbraith, M.N., D.H.S. Horn, E.J. Middleton, and J.A. Thomson. (1970) The biosynthesis of crustecdysone in the blowfly Calliphora stygia. Chem. Commun. 3: 179-80.

Goodnight, K.C. and H.W. Kircher. (1971) Metabolism of lathosterol by Drosophila pachea. Lipids 6: 166-69.

Heed, W.B. and H.W. Kircher. (1965) Unique sterol in the ecology and nutrition of Drosophila pachea. Science 149: 758-61.

Herbert, E.W., Jr., J.A. Svoboda, M.J. Thompson, and H. Shimanuki. (1980) Sterol utilization in honey bees fed a synthetic diet: effects on brood rearing. J. Insect Physiol. 26: 287-89.

Hobson, R.P. (1935) On a fat soluble growth factor required by blowfly larvae. II. Identity of the growth factor with cholesterol. Biochem. J. 29: 2023-26.

Hugel, M.F. (1962) Etude des quelques constituents du pollen. (A study of some components of pollen). Annls. Abeille 5: 97-133.

Kaplanis, J.N., S.R. Dutky, W.E. Robbins, M.J. Thompson, E.L. Lindquist, D.H.S. Horn, and M.N. Galbraith. (1975) Makisterone A: A 28-carbon hexahydroxy molting hormone from the embryo of the milkweed bug. Science 190: 681-82.

Kaplanis, J.N., R.E. Monroe, W.E. Robbins, and S.J. Louloudes. (1963) The fate of dietary ^{3}H-ß-sitosterol in the adult house fly. Ann. Entomol. Soc. Am. 56: 198-201.

Kaplanis, J.N., W.E. Robbins, R.E. Munroe, T.J. Shortino, and M.J. Thompson. (1965) The utilization and fate of ß-sitosterol in the larva of the housefly, Musca domestica L. J. Insect Physiol. 11: 251-58.

Karlson, P. and H. Hoffmeister. (1963) Zur Biogenese des Ecdysons, I Umwandlung von Cholesterin in Ecdyson. Z. Physiol. Chem. 331: 298-300.

Robbins, W.E., R.C. Dutky, R.E. Monroe, and J.N. Kaplanis. (1962) The metabolism of ^{3}H-ß-sitosterol by the German cockroach. Ann. Entomol. Soc. Am. 55: 102-04.

Robbins, W.E., J.N. Kaplanis, J.A. Svoboda, and M.J. Thompson. (1971) Steroid metabolism in insects. Ann. Rev. Entomol. 16: 53-72.

Standifer, L.N., M. Davys, and M. Barbier. 1968. Pollen sterols--a mass spectrographic survey. Phytochemistry 7: 1361-65.

Svoboda, J.A., S.R. Dutky, W.E. Robbins, and J.N. Kaplanis. (1977) Sterol composition and phytosterol utilization and metabolism in the milkweed bug. Lipids 12: 318-21.

Svoboda, J.A., E.W. Herbert, Jr., M.J. Thompson, and H. Shimanuki. 1981. The fate of radiolabeled C_{28} and C_{29} phytosterols in the honey bee. J. Insect Physiol. 27: 183-88.

Svoboda, J.A., J.N. Kaplanis, W.E. Robbins, and M.J. Thompson. (1975) Recent developments in insect steroid metabolism. Ann. Rev. Entomol. 20: 205-20.

Svoboda, J.A., A.M.G. Nair, N. Agarwal, H.C. Agarwal, and H.C. Robbins. (1979) The sterols of the khapra beetle, Trogoderma granarium Everts. Experientia 35: 1454-55.

Svoboda, J.A., A.M.G. Nair, N. Argarwal, and W.E. Robbins. (1980) Lack of conversion of C_{29}-phytosterols to cholesterol in the khapra beetle, Trogoderma granarium Everts. Experientia 36: 1029-1030.

Svoboda, J.A. and W.E. Robbins. (1971) The inhibitive effects of azasterols on sterol metabolism and growth and development in insects with special reference to the tobacco hornworm. Lipids 6: 113-19.

Svoboda, J.A., and W.E. Robbins. (1979) Comparison of sterols from a phytophagous and predacious species of the family Coccinellidae. Experientia 35: 186-87.

Svoboda, J.A., W.E. Robbins, C.F. Cohen, and T.J. Shortino. (1972) Phytosterol utilization and metabolism in insects: recent studies with Tribolium confusum. In Insect and Mite Nutrition, ed. J.G. Rodriguez, 505-16. Amsterdam: North-Holland. 702 pp.

Svoboda, J.A., M.J. Thompson, and W.E. Robbins. (1967) Desmosterol, an intermediate in dealkylation of ß-sitosterol in the tobacco hornworm. Life Sci. 6: 395-404.

Svoboda, J.A., M.J. Thompson, T.C. Elden, and W.E. Robbins. (1974) Unusual composition of sterols in a phytophagous insect, Mexican bean beetle reared on soybean plants. Lipids 9: 752-55.

Svoboda, J.A., M.J. Thompson, E.W. Herbert, Jr., and H. Shimanuki. (1980a) Sterol utilization in honey bees fed a synthetic diet: analysis of prepupal sterols. J. Insect Physiol. 26: 291-94.

Svoboda, J.A., M.J. Thompson, W.E. Robbins, and T.C. Elden. (1975a) Unique pathways of sterol metabolism in the Mexican bean beetle, a plant-feeding insect. Lipids 10: 524-27.

Svoboda, J.A., M.J. Thompson, W.E. Robbins, and J.N. Kaplanis. (1978) Insect steroid metabolism. Lipids 13: 742-53.

Svoboda, J.A., M.J. Thompson, and T.J. Shortino. Unpublished results.

16

Thompson, M.J., S.J. Louloudes, W.E. Robbins, J.A. Waters, J.A. Steele, and E. Mosettig. 1963. The identity of the major sterol from houseflies reared by the CSMA procedure. J. Insect Physiol. 9: 615-22.

2
Sterol Metabolism of the Silkworm *Bombyx mori*

Masuo Morisaki, Yoshinori Fujimoto,
Akihiro Takasu, Yoko Isaka,
and Nobuo Ikekawa*

INTRODUCTION

Insects are incapable of de novo sterol synthesis and therefore require a dietary or exogenous source of sterol for their normal growth and development. The sterol requirement of insects is, in most cases, satisfied by cholesterol. This is the principal sterol in insects and serves the role of structural component of cell membranes and as biogenetic precursor of the moulting hormone ecdysone. In phytophagous insects however, the requirement can be fulfilled also with plant sterols e.g. sitosterol, stigmasterol and campesterol, since these C-24 alkylated sterols are metabolically converted to cholesterol. Thus, dealkylation of phytosterol is an important metabolic step in many insects. The biochemical mechanism of conversion of plant sterols to cholesterol has been investigated principally by Svoboda's group utilizing tobacco hornworms. They have identified several intermediates and proposed the pathway summarized in Fig. 1 (Thompson et al., 1972).

However, the precise carbon-24,28 bond cleavage mechanism of plant sterols had not been clarified as indicated by dotted lines in Fig. 1. Since the silkworm <u>Bombyx mori</u> was previously shown to convert [^{3}H]-sitosterol to cholesterol (Ikekawa et al., 1966) and a satisfactory agar-base artificial diet is available (Ito et al., 1964), we have studied phytosterol dealkylation using this phytophagous insect.

---

* Laboratory of Chemistry for Natural Products, Tokyo Institute of Technology, Nagatsuta, Midori-ku, Yokohama 227, Japan.

FUCOSTEROL 24,28-EPOXIDE AS AN ITERMEDIATE OF SITOSTEROL
DEALKYLATION

A brief treatment of fucosterol epoxide with Lewis
acid, e.g. boron trifluoride etherate, afforded among
other compounds desmosterol at 35% yield (Ikekawa et
al., 1971; Ohtaka et al., 1973). It was hypothesized
that a similar reaction would also occur in the insect.
This was subsequently verified on the basis of the following evidence. When [^{3}H]-fucosterol epoxide was
ingested by silkworm larvae, it was converted to cholesterol with a high yield, and also the tritium of [^{3}]-
fucosterol was incorporated into fucosterol epoxide
(Morisaki et al., 1972). Tritium of [25-^{3}H]-sitosterol
migrated to the C-24 position during its conversion to
desmosterol (Fujimoto et al., 1974a). Fucosterol epoxide
completely satisfied the silkworm sterol requirement
(Morisaki et al., 1974). Fucosterol epoxide was isolated

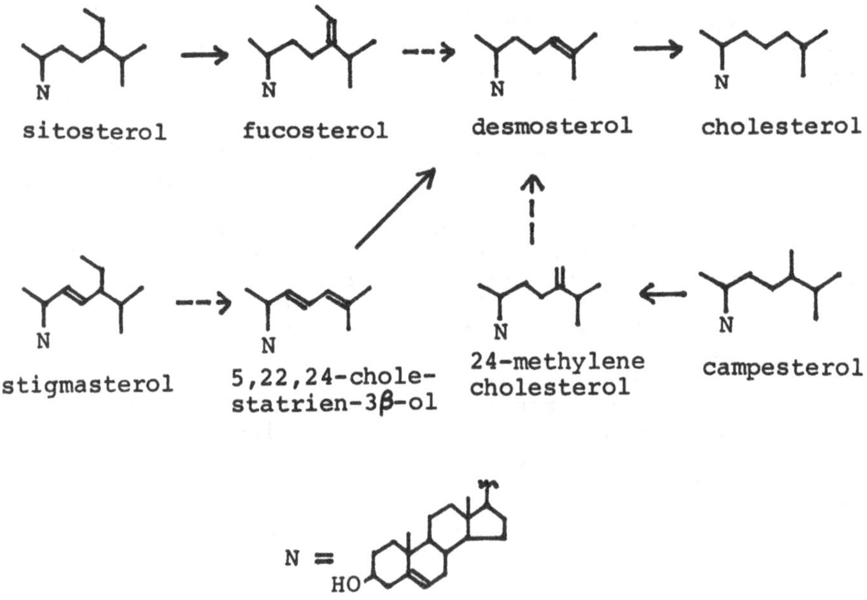

FIGURE 1. Conversion of phytosterols to cholesterol in
insect (Thompson, et al., 1972)

and definitely identified from B. mori larvae reared on
mulberry leaves, the natural diet of this insect (Ikekawa
et al., 1980). Thus, the key intermediary role of fucosterol epoxide in sitosterol dealkylation in B. mori
seems established (Fig. 2). Other insects seem also to
dealkylate sitosterol in a similar manner (Randall et
al., 1972; Allis et al., 1973). A step-wise dealkylation
route, reminiscent of the reverse of phytosterol biosynthesis (i.e. through isofucosterol and 24-methylene cholesterol), seems to be of minor importance.

Attention was then turned to the stereochemistry of
the epoxide. In order to know which of the two stereoisomers of fucosterol 24,28-epoxide, the 24R, 28R or the
24S, 28S (Fig. 2), is the true intermediate, they were
chemically synthesized (Chen et al., 1975; Fujimoto et
al., 1980a). When the supernatant obtained from homogen-

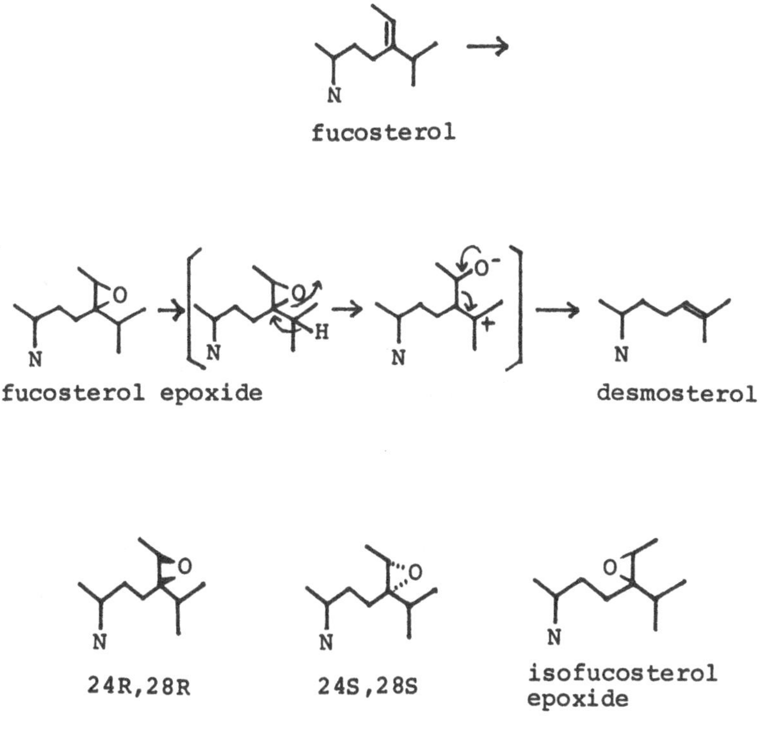

fucosterol

fucosterol epoxide

desmosterol

24R,28R

24S,28S

isofucosterol
epoxide

FIGURE 2. Fucosterol 24,28-epoxide as the intermediate
of conversion of fucosterol to desmosterol in Bombyx
mori. (N as in Fig. 1).

ates (Awata et al., 1975) of Bombyx mori gut was incubated separately with the [^{3}H]-24R,28R-epoxide and the 24S,28S-epoxide, only the former was effectively converted to desmosterol (Chen et al., 1975). However, the subsequent in vivo experiments and the more precise in vitro studies (Fujimoto et al., 1980b) demonstrated no absolute stereospecificity both in the formation of the epoxide from fucosterol and in its conversion to desmosterol, suggesting both of the epoxides could be intermediates. This supposition was substantiated when the two epoxides were found to equally satisfy the B. mori sterol requirement and the sterol profiles of the insects reared on them showed no significant difference (Fujimoto et al., 1980b). Moreover, fifth instar larvae reared on mulberry leaves showed the presence of both of the two stereoisomers (Ikekawa et al., 1980). The unexpected low stereospecificity observed with fucosterol epoxide appears to be in line with the facts that both fucosterol and isofucosterol (the geometrical isomer of fucosterol) were identified in B. mori (Morisaki et al., 1981) and no discrimination occurs in the nutritional effects of sitosterol and clionasterol (the C-24 epimer of sitosterol) (Fujimoto et al., unpublished). We are now inclined to consider that dealkylation of 24-ethylcholesterol in B. mori would proceed with a low degree of stereospecificity.

Although an intermediary role of isofucosterol epoxide (Fig. 2) appears to be of slight importance in B. mori, in view of its poor nutritional effect (Fujimoto et al. 1980b) and absence from the larvae (Ikekawa et al. 1980), this may not be the case in Tenebrio molitor (Nicotra et al. 1981).

STRUCTURAL ANALOGS OF SITOSTEROL DEALKYLATION INTERMEDIATES AS POSSIBLE INHIBITORS OF STEROL METABOLISM

Inhibition of sterol metabolism has proven valuable for studying metabolism of plant sterols in insects (Thompson et al., 1972). For example, 22,25-diazacholesterol is an inhibitor of Δ^{24} sterol reductase and was used for identification of desmosterol and cholesta-5,22,24-trien-3ß-ol as intermediates of plant sterol dealkylation. We postulated that structural analogs of sitosterol dealkylation intermediates would be potent and specific inhibitors of sterol metabolism of Bombyx mori. As possible candidates, 24,28-iminofucosterol(imine), stigmasta-5,24(28)-trien-3ß-ol (allene I) and cholesta-5,23,24-trien-3ß-ol (allene II) were prepared (Fujimoto et al., 1974b; Morisaki et al. 1975; Fujimoto et al., 1975). As evident from Fig. 3, these compounds can be regarded as the structural analogs of fucosterol epoxide, fucosterol and desmosterol, respectively.

When the imine or the allene II were administered in the silkworm diet in combination with sitosterol or cholesterol, the growth and development of B. mori were markedly retarded. Insect sterol analysis revealed that the imine is a sitosterol dealkylation inhibitor, and brings about accumulation of the unchanged sitosterol (Fujimoto et al., 1974b). The one likely target of the imine inhibition was expected to be the step of conversion of fucosterol epoxide to desmosterol, and this was verified with in vitro experiments where the imine, at the same level with the substrate [^{3}H]-fucosterol epox-

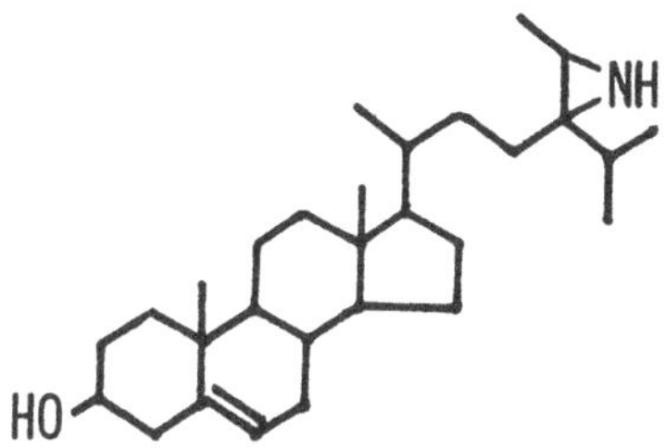

imine 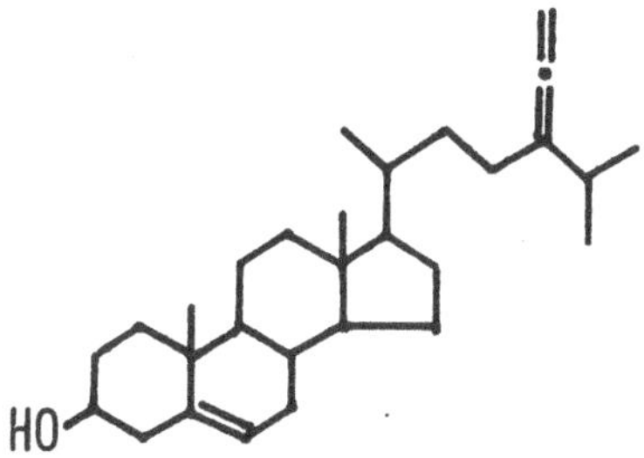   allene I

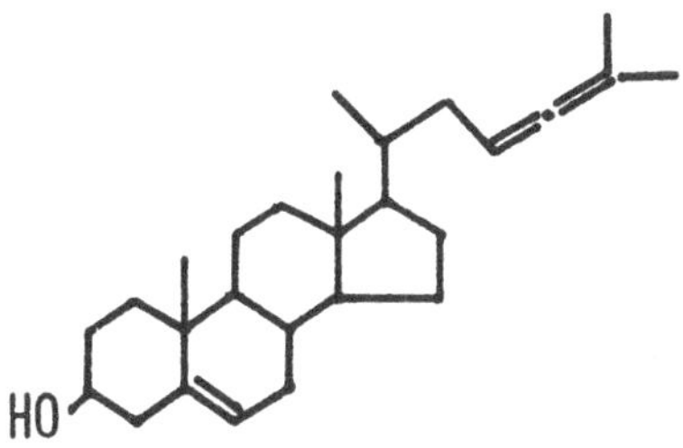

allene II

FIGURE 3. Structural analogs of sitosterol dealkylation intermediates.

ide, completely blocked the transformation into desmosterol (Awata et al., 1975). However, the imine may exert
its effect other than solely by limitation of desmoster-
ol/cholesterol formation, because cholesterol used as
the sole dietary sterol was unable to prevent the imine
inhibitory effect. In contrast, the allene II seemed to
exert little effect on sitosterol dealkylation because
the sterol components in silkworm fed the allene II in
combination with sitosterol were essentially not different from those of the controls (Awata et al., 1976).

Allene I has more interesting properties. When B.
mori was reared on fucosterol epoxide, desmosterol or
cholesterol, the addition of allene I to the diets, at
the same concentration as nutritional sterol, caused no
significant inhibition of insect growth and development.
In contrast, when the nutritional sterol was replaced
with sitosterol or fucosterol, marked growth retardation
was observed. In agreement with these results, the
sterol component of the insect was not appreciably influenced by added allene I, when the sterol source was fucosterol epoxide, desmosterol or cholesterol, whereas
allene I induced the accumulation of sitosterol or fucosterol when these sterols were the dietary sterol source.
These results strongly suggest that allene I would be a
highly specific inhibitor for the steps involving fucosterol, that is for the conversion of sitosterol to fucosterol and/or fucosterol to fucosterol epoxide (Awata et
al., 1976).

STRUCTURAL REQUIREMENT OF STEROL SIDE CHAIN FOR SILKWORM
GROWTH AND DEVELOPMENT

B. mori can grow and develop on a semi-synthetic diet
containing cholesterol or phytosterols at 0.1-0.5% (Ito
et al., 1964). It was hypothesized that possible intermediates of phytosterol dealkylation would also satisfy
the sterol requirement of B. mori (Morisaki et al.,
1974). Several sterols in these pathways were tested for
their nutritional effect and were classified as "effective", if they behaved in the same manner as cholesterol
or phytosterols, "partially effective", if B. mori sur-
vived through the early instars but with a slower growing
rate, and "ineffective", if all larvae died in the first
instar (Fig. 4). As expected, fucosterol, fucosterol
epoxide and desmosterol belonged to the effective group
of sterols, whereas 28-oxo-, 24,28dihydroxy- and 24-
hydroxy-28-oxositosterol and 24-oxo- and 24-hydroxy-
cholesterol were all placed in the ineffective group of
sterols. These facts probably exclude a dealkylation
route analogous to the side chain cleavage of cholesterol and pregnenolone in vertebrates.

By the use of this methodology, we attempted to get

Effective sterol

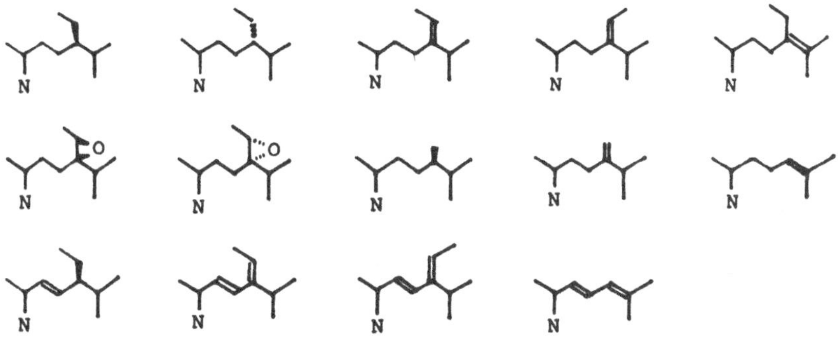

Partially effective sterol

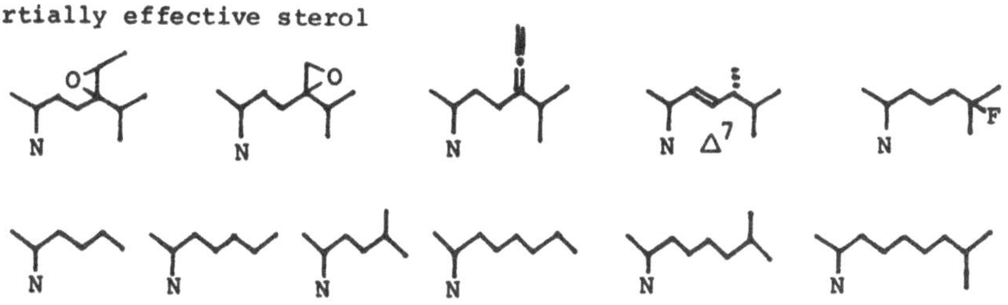

Ineffective sterol

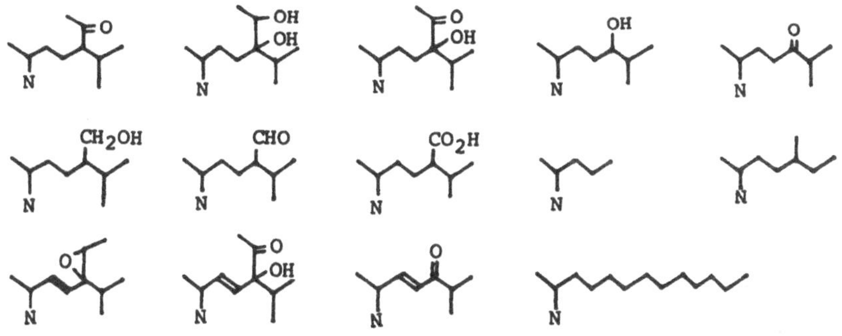

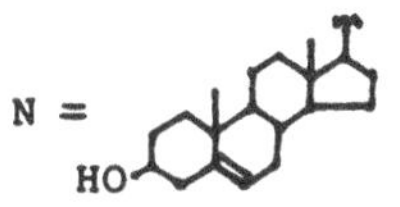

FIGURE 4. Sterol classification based on their nutritional effect on <u>Bombyx</u> <u>mori</u>

preliminary information about the intermediates of campesterol and stigmasterol dealkylation. In spite of the
complete satifaction of the sterol requirement with 24-
methylene cholesterol, this epoxide could only inade-
quately support the growth of B. mori. Accordingly 24-
methylene cholesterol may be demethylated without passing
through the epoxide, and the dealkylation mechanism of
campesterol would be different from that of sitosterol.
An alternative route might be an oxidative one passing
through the 28-alcohol, 28-aldehyde and/or 28-carboxylic
acid, but all of these compounds were classified as ineffective sterols (Fig. 4) (Takasu et al., unpublished).
As for dealkylation of stigmasterol, the three trienes,
i.e. (24E)- and (24Z)-stigmasta-5,22,24(28)-trien-3ß-ol
and cholesta-5,22,24-trien-3ß-ol were found to be effective sterols, whereas the 24,28-epoxide derived from
stigmastatrienes was ineffective. In addition, 28-oxo-
stigmasta-5,22-diene-3ß,24-diol and cholesta-5,22-dien-
24-on-3ß-ol were also ineffective nutrients. These
results are incompatible either with the epoxide route
or the oxidative pathway of stigmasterol dealkylation
(Takasu et al., unpublished). It should be noted, however, that most of the ineffective sterols described
above are relatively hydrophilic in nature and may not
pass the cell membrane barrier to the site of dealkylation. Furthermore, when these sterols are ingested in
such large amounts as used for nutritional experiments
they may deleteriously affect the normal physiology of
insects. Therefore, further investigations are needed
before coming to a definite conclusion.

Included also in Fig. 4 are several cholesterol analogs with the side chain modified in carbon chain length
and branching pattern (Morisaki et al., 1980). These
compounds were tested for their nutritional effect in
order to have an insight into the relationship between
biological function and the side chain structure of cholesterol. Most of the larvae fed the longest (C_{32}) or
the shortest (C_{24}) side chains died during the first
instar. The other cholesterol analogs tested were partially effective sterols; the growth rate decreased the
more the structure deviated from cholesterol (Isaka et
al., 1981). Insect sterol analysis revealed that the
respective dietary sterols were almost unchanged in B.
mori, comprising more than 95% of the total tissue
sterol. This suggests that the major membrane sterol
would be the particular dietary sterol itself, and that
the moulting hormone might not be the usual native ecdysone (derived from cholesterol via 7-dehydrocholesterol)
but a modified one. It may reasonably be anticipated
that such modification of insect membrane sterols and
moulting hormones would be only partially effective for
normal biological functions.

Acknowledgement. We thank Dr. H. Maekawa for Bombx
mori eggs. We are also grateful to Drs. N. Awata, H.
Ohtaka, Y. Horie and S. Nakasone for their contribution
in the earlier stage of this work. This work was supported by research grants from the Ministry of Education,
Japan.

REFERENCES

Allais, J.P., Alcaide, A. and Barbier, M. (1973) Fuco-
    sterol-24,28 epoxide and 28-oxo-ß-stiosterol as pos-
    sible intermediates in the conversion of ß-sitosterol
    into cholesterol in the locust Locusta migratoria L.
    Experientia, 29: 944-945.
Awata, N., Morisaki, M. and Ikekawa, N. (1975) Carobon-
    carbon bond cleavage of fucosterol-24,28-oxide by
    cell-free extracts of silkworm Bombyx mori. Biochem.
    Biophys. Res. Comm., 64: 157-161.
Awata, N., Morisaki, M., Fujimoto, Y. and Ikekawa, N.
    (1976) Inhibitory effects of steroidal allenes on
    growth, development and steroid metabolism of the
    silkworm Bombyx mori. J. Insect Physiol., 22: 403-
    408.
Chen, S. M.L., Nakanishi, K., Awata, N., Morisaki, M.,
    Ikekawa, N., and Shimizu, Y. (1975) Stereospecific-
    ity in the conversion of fucosterol 24,28-epoxide to
    desmosterol in the silkworm, Bombyx mori. J. Amer.
    Chem. Soc., 97: 5297-5299.
Fujimoto, Y., Awata, N., Morisaki, M. and Ikekawa, N.
    (1974a) Migration of C-25 hydrogen of sitosterol to
    C-24 during the conversion into desmosterol in the
    silkworm Bombyx mori. Tetrahedron Lett., 4335-4338.
Fujimoto, Y., Morisaki, M. Ikekawa, N., Horie, Y. and
    Nakasone, S. (1974b) Synthesis of 24,28-iminofuco-
    sterol and its inhibitory effects on growth and
    steroid metabolism in the silkworm, Bombyx mori.
    Steroids, 24, 367-375.
Fujimoto, Y., Morisaki, M., and Ikekawa, N. (1975)
    Studies on Steroids. Part XXIX. Synthesis of allen-
    ic analogues of fucosterol and desmosterol. J. Chem.
    Soc. Perkin I: 2302-2307.
Fujimoto, Y., Murakami, K. and Ikekawa, N. (1980a) Syn-
    thesis of (24R,28R)- and (24S-28S)-fucosterol epox-
    ides. Revision of C-24,28 configurations. J. Org.
    Chem., 45, 566-569.
Fujimoto, Y., Morisaki, M. and Ikekawa, N. (1980b)
    Sterochemical importance of fucosterol epoxide in
    the conversion of sitosterol into cholesterol in the
    silkworm Bombyx mori. Biochemistry, 19, 1065-1069.
Ikekawa, N., Suzuki, M., Kobayashi, M. and Tsuda, K.
    (1966) Studies on the sterol of Bombyx mori L. IV.
    Conversion of the sterol in the silkworm. Chem.

Pharm. Bull., 14, 834-836.

Ikekawa, N., Morisaki, M., Ohotaka, H. and Chiyoda, Y. (1971) Reaction of fucosterol 24,28-epoxide with boron trifluoride etherate. Chem. Comm., 1498.

Ikekawa, N., Fujimoto, Y., Takasu, A., and Morisaki, M. (1980) Isolation of fucosterol epoxide from larvae of the silkworm, Bombyx mori. J. Chem. Soc. Chem. Comm., 709-711.

Isaka, Y., Morisaki, M. and Ikekawa, N. (1981) Structural requirement of sterol side chain for the silkworm growth and development. Steroids, in press.

Ito, T., Kawashima, K., Nakahara, M., Nakanishi, K. and Terahara, A. (1964) Effect of sterols on feeding and nutrition of the silkworm, Bombyx mori. J. Ins. Physiol., 10, 225-238.

Morisaki, M., Ohotaka, H., Okubayashi, M., Ikekawa, N., Horie, Y. and Nakasone, S. (1972) Fucosterol-24,28-epoxide, as a probable intermediate in the conversion of ß- sitosterol to cholesterol in the silkworm. J. Chem. Soc. Chem. Comm., 1275-1276.

Morisaki, M., Ohotaka, H., Awata, N., Ikekawa, N., Horie, Y. and Nakasone, S. (1974) Nutritional effect of possible intermediates of phytosterol dealkylation in the silkworm Bombyx mori. Steroids, 24, 1655-176.

Morisaki, M., Awata, N., Fujimoto, Y. and Ikekawa, N. (1975) Steroidal allenes as inhibitors of sterol biosynthesis. J. Chem. Soc. Chem. Comm., 362-363.

Morisaki, M., Shibata, M., Duque, C., Imamura, N. and Ikekawa, N. (1980) Studies on steroids. LXIII. Synthesis of cholesterol analogs with a modified side chain. Chem. Pharm. Bull., 28, 606-611.

Morisaki, M., Ying, B. and Ikekawa, N. (1981) Identification of both fucosterol and isofucosterol in the silkworm, Bombyx mori. Experientia, 37, 336-337.

Nicotra, F., Pizzi, P., Ronchetti, F. and Russo, G. (1981) Substrate specificity in the metabolism of fucosterol and isofucosterol 24,28-epoxides in Tenebrio molitor. J. Chem. Soc. Perkin I, 480-483.

Ohotaka, H., Morisaki, M. and Ikekawa, N. (1973) Reaction of 24,28-epoxides of sterol side chain with boron triflouride etherate. J. Org. Chem., 38, 1688-1691.

Randall, P.J., Lloyd-Jones, J.G., Cook, I.F., Rees, H.H. and Goodwin, T.W. (1972) The fate of the C-25 hydrogen of 28-isofucosterol during conversion into cholesterol in the insect Tenebrio molitor. J. Chem. Soc. Chem. Comm., 1296-1298.

Thompson, M.J., Svoboda, J.A., Kaplanis, J.N. and Robbins, W.E. (1972) Metabolic pathway of steroids in insects. Proc. R. Soc. Lond. B. 180, 203-221.

3
Lipid Interdependencies Between *Xyleborus* Ambrosia Beetles and Their Ectosymbiotic Microbes

K.D.P. Rao, D. M. Norris,
and H. M. Chu*

Many species of insects are known to have symbiotic microbes (e.g., Batra, 1979). Such symbiotes, especially ectosymbiotes, commonly interface the insect nutritionally and physiologically with some primary energy substrate (e.g., lignocellulose) (Norris, 1979). Thus, the insect obtains its required nutrition by cultivating symbiotes on wood, and then eating the microbial mass. During the coevolution of such highly specialized symbiosis unusual nutritional requirements might be expected to arise in the insect, microbes or both. This chapter details our knowledge of nutritional and physiological lipid interdependencies which exist between the scolytid ambrosia beetles, Xyleborus spp. (especially X. ferrugineus, and ectosymbiotic microbes.

Ectosymbiotes especially provide three types of important nutritive lipids, sterols, phospholipids and fatty acids, to Xyleborus beetles. Sterols serve as precursors for essential steroid metabolites, such as ecdysteroid hormones, and as structural components of cells. Phospholipids constitute varying proportions of body fat and cell membranes. Fatty acids are important as energy sources and as components of phospholipids and glycerides.

STEROIDAL INTERDEPENDENCIES

Insects per se are unable to synthesize sterols, and thus require dietary sterol. Most studied insects can utilize cholesterol as a sole dietary sterol source; however, X. ferrugineus requries a dietary Δ^7-sterol for normal growth, development, reproduction and longevity (Chu et al., 1970; Chu and Norris, 1970; Norris, 1972;

* Department of Entomology, University of Wisconsin, Madison, WI, 53706, USA.

27

Norris and Moore, 1980). This obligatory nutrient is
supplied to the insect in the form of ergosterol by ectosymbiotic fungi, i.e., <u>Fusarium</u> <u>solani</u>, <u>Cephalosporium</u>
sp. and <u>Graphium</u> sp. (Kok and Norris, 1973). Thus, the
insect's growth, development, reproduction and longevity
are especially controlled biochemically by its mutualistic ectosymbiotes through steroid metabolism.

Adult Female

Our studies of specific effects of various dietary
sterols on <u>Xyleborus</u> adult females have revealed that the
maturation of oocytes is affected before locomotory vigor
or lifespan (Norris and Moore, 1980). At age 16.4 weeks,
only 50% of the female adults which received only cholesterol as dietary sterol were still ovipositing; whereas,
50% of those which received only ergosterol as dietary
sterol were still ovipositing at 27.4 weeks. Comparative
values for the effects of these two dietary sterols on
locomotory vigor (i.e., tunneling in the diet) were: cholesterol, 50% tunneling at 20.3 weeks of age; and ergosterol, 50% at 30.3 weeks. Effects on female longevity
were: cholesterol, 50% survival at 19.5 weeks of age;

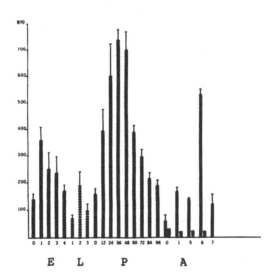

FIGURE 1. Ecdysteroid titers (vertical scale: pg/mg
body wgt.) during development of <u>X</u>. <u>ferrugineus</u>. E:
0 = newly laid eggs; 1-4 = stages of embryonic development. L: 1, 2 and 3 indicate larval instars. P: pupal
age in hours. A: adult age in days; uniformly striped
bars, females; other bars, males.

and ergosterol, 50% at 30.5 weeks. The effects on all three studied parameters, i.e., reproduction (oocyte maturation and oviposition), locomotory vigor and longevity, of not receiving the Δ^7-sterol nomally provided by symbiotes probably involve both ecdysteroid-hormone and tissue-component-sterol "pools" in the female (Norris and Moore, 1980).

The specific involvement of each of these two steroidal pools in the ovarian maturation of Xyleborus already has been demonstrated experimentally. The titer of ecdysteroids (determined by the radioimmunoassay (RIA) of Bollenbacher et al., 1975) in the body of the fertile female surges to a peak during day 6 (Fig. 1) of her ovarian cycle (Rao et al., 1982). If this marked increase in hormone titer is prevented experimentally (Bridges and Norris, 1977; Chu et al., 1982a), then oocyte maturation terminates as with the aging beetle. Thus, the ecdysteroid-hormone pool is especially involved in regulating reproduction (e.g., oocyte maturation) in the female beetle. In addition, in the very aged beetle (e.g., 30-weeks old on dietary cholesterol) there also is extensive ultrastructural evidence of the deterioration of intra-cellular membranes in the ovary. This is attributable to receiving only cholesterol, and not a Δ^7-sterol, as dietary sterol (Chu et al., 1982b).

Adult Male

The male (haploid) adult also depends on the ectosymbiotes for dietary sterol. The specific role of ecdysteroid-hormone versus tissue-component-sterol "pools" in this sex remains unknown. However, it is known that the basal titer of free ecdysteroids in males seems to be about one-half of that in the diploid female (Fig. 1). Contrary to the situation in the fertile female, the titer in males remains fairly constant during the vigorous period of its life.

Immature Stages: Ecdysteroid Contents

Ecdysteroids in immature stages of Xyleborus must originate either from the mother beetle and be transferred transovarially to progeny, or from the metabolism of dietary Δ^7-sterol provided by ectosymbiotic microbes (Norris, 1972; Kok, 1979). Ecdysteroid titers in the whole body of various immature stages are given in Fig. 1. Titers were measured in eggs of five age groups (stages): 0) newly laid eggs, 1) embryos that had mostly undergone gastrulation, 2) embryos that were between germ-band contraction and definitive dorsal closure, 3) embryos that had undergone initial organogenesis, and group 4) embryos that were mature but before eclosion.

The content of ecdysteroids in the freshly laid eggs was 142 $\pm$ 21 pg/mg egg wt. The titer of these hormones peaked at 362 $\pm$ 47 pg/mg egg wt in Stage 1 embryos and then declined gradually. By the time the embryos completed organogenesis and were ready for eclosion, the titer had declined to 174 $\pm$ 21 pg/mg egg wt.

The increase in RIA-positive compounds in Stage 1 embryos could be due to either de novo synthesis, or the unmasking of previously conjugated forms. The freeing of ecdysteroids from conjugated forms is the most likely source of the increased titer of these hormones. With either of these situations, the outcome is the availability of significantly higher quantities of ecdysteroids to embryos which have finished gastrulation. The presence of such increased titers implies major roles for these hormones during post gastrulation embryonic development. The pattern of ecdysteroid titers during the embryonic development of Xyleborus somewhat resembles that in Schistocerca (Gande and Morgan, 1979). Other insects known to have ecdysteroids in the embryos include Galleria (Hsiao and Hsiao, 1979), Nauphoeta (Imboden et al., 1978) and Locusta (Lagueux et al., 1977).

The titer of ecdysteroids declined to 72 $\pm$ 14 pg/mg body wt in mid first-instar larvae (Fig. 1). During the mid second-larval stadium the titer peaked at 193 $\pm$ 49 pg/mg. However, there was no significant change in the ecdysteroid content during the third larval instar; the average titer was at 97 $\pm$ 27 pg/mg body wt. The pattern of ecdysteroid titers during the larval stages of Xyleborus resembled those of Leptinotarsa (Hsiao et al., 1976) and Drosophila (Borst et al., 1974). The ecdysteroid peak in Xyleborus larvae occurred significantly earlier than in larvae of most studied insect species. The highest concentration of the ecdysteroids in the larvae was much less than was found in the pupae (Fig. 1). This would support the conclusion that larval-pupal transformation requires lower levels of these hormones than does pupal-adult transformation.

Ecdysteroid titer in the newly formed pupa was 161 pg/mg (Fig. 1). As the pupa initially aged, the ecdysteroid content increased; this increase was most significant between 0-24 hr. After peaking (743 pg/mg body wt) at 36 hr in pupal development the titer declined. The decline between 36-48 hr was gradual, but it became precipitous between 48-60 hr. Pharate adult development starts around 60 hr and the ecdysteroid titer then was 391 pg/mg body wt. From the onset of pharate adult development until the emergence of the adult beetle a gradual decline in the ecdysteroid titer was observed. By having a single ecdysteroid peak during pupal and pharate adult development Xyleborus resembles Tenebrio (Delbecque et al., 1978), Heliothis (Holman and Meola,

1978), <u>Leptinotarsa</u> (Hsiao et al., 1976) and <u>Sarcophaga</u> (Ohtaki and Takahashi, 1972); and differs from <u>Galleria</u> (Bollenbacher et al., 1978) and <u>Bombyx</u> (Hanaoka and Ohnishi, 1974). In comparison to most insects the peaking of ecdysteroid titer during the pupal stage of <u>Xyleborus</u> is slower and more gradual. This situation suggests a steady synthesis (and/or release) from the organ(s) concerned. Another interesting observation in <u>Xyleborus</u> is that the ecdysteroid levels during pharate adult development did not fall to threshold (i.e., zero) level and RIA-positive compounds were detected even in post-eclosion females. While this finding agrees with those of Delbecque et al. (1978) in <u>Tenebrio</u> it is in marked contrast with the observations of Hsiao et al. (1976) in the coleopteran <u>Leptinotarsa</u>.

The qualitative composition of ecdysteroids changes during the metamorphosis of insects, and high performance liquid chromatography (HPLC) of whole body extracts of <u>Xyleborus</u> pupae at three stages of development revealed quantitative changes between two major ecdysteroids, ecdysone and 20-hydroxyecdysone (Rao <u>et al</u>., 1981). In the first phase (Fig. 2), the ratio of ecdysone to 20-hydroxyecdysone was about 3:1. A progressive shift in this ratio was seen during phase 2 of pupal development (Fig. 3). During the latter half of the pupal development, which includes mainly pharate adult development, 20-hy-

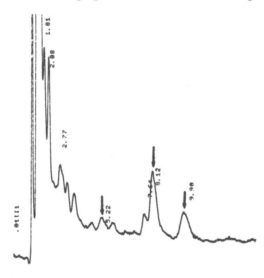

FIGURE 2. HPLC separation of RIA-positive compounds from phase 1 pupa (0=24 hr) of X. <u>ferrugineus</u>. 20-hydroxyecdysone (5.22), unknown-1 (7.64), unknown-11 (8.12) and ecdysone (9.98).

droxyecdysone became dominant (Fig. 4). A high ecdysone titer in the initial pupal stage followed by a high concentration of 20-hydroxyecdysone during the latter period of pupal development may mean that 20-hydroxyecdysone is the morphogenetic hormone in Xyleborus. Besides ecdysone and 20-hydroxyecdysone, a smaller peak (Unknown-1) at RT 7.64 was detected during phase 1 of pupal development (Fig. 2) and a major peak (Unknown-11) was observed throughout pupal and pharate adult development. We believe that these two unknown peaks are ecdysteroids. Changing ecdysteroid titers during pupal development also were shown in Heliothis (Holman and Meola, 1978) and Drosophila, where only 20-hydroxyecdysone was detected in the methanolic extracts (Borst et al., 1974).

ORGANS OF ECDYSTEROID SYNTHESIS

The synthesis of ecdysteroids by prothoracic glands has been established in various insect species (Gilbert and King, 1973). Detailed ultrastructural studies of pupal Xyleborus prothoracic glands during low (0-hr pupal development) and high (48-hr pupal development) ecdysteroid-synthesizing periods revealed that at 0-hr old (Fig. 5) the gland cells are very compact, the nucleus occupies a large portion of the cell, variously shaped and sized mitochondria with a dense matrix are present and distinct cristae are spread throughout the cytoplasm; whereas, at 48-hr old (Fig. 6) gland cells are tightly arrayed and compact, they have a large nucleus and relatively little cytoplasm, the vacuolar system is quite distinctive and the number of mitochondria is markedly reduced (Chu et al., 1980). The matrix of such cells is less dense than that of the 0-hr-old prothoracic glands, free ribosomes and glycogen particles are present and more or less parallel microtubules are abundant. The lysosomelike structures found in the prothoracic gland cells of 0-hr-old pupae are not present at this 48-hr-old stage.

The observed ultrastructural parameters do not strongly suggest steroidal hormone synthesis or secretion by the glands. Glitho et al. (1979) showed synthesis of ecdysteroids by organs other than prothoracic glands in Tenebrio. This situation was further confirmed by Delbecque et al. (1978) who showed substantial production of ecdysteroids in isolated pupal abdomens of male and female Tenebrio. Hsiao and Hsiao (1975 and Hsiao et al. (1976) further showed that isolated abdomens of post-feeding larvae of Leptinotarsa were able to produce ecdysteroids. Oenocytes also were suggested as good candidates for extra-prothoracic gland synthesis of ecdysteroids (Delachambre, 1980; Romer et al., 1974).

The most interesting observation to emerge from our

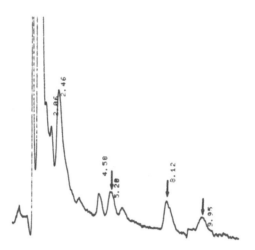

FIGURE 3. HPLC separation of RIA-positive compounds from
phase 2 pupa (24-48 hr) of X. ferrugineus. 20-hydroxy-
ecdysone (5.28), unknown-11 (8.12) and ecdysone (9.95).

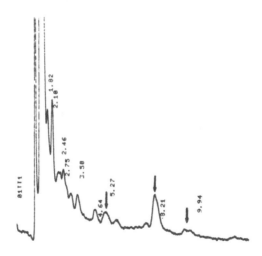

FIGURE 4. HPLC separation of RIA-positive compounds from
phase 3 pupa (48-96 hr) of X. ferrugineus. 20-hydroxy-
ecdysone (5.27), unknown-11 (8.21) and ecdysone (9.94).

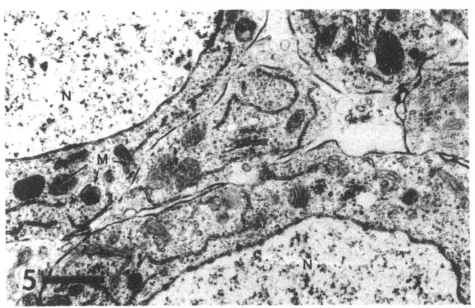

FIGURE 5. Prothoracic gland cells of 0-hr-old female pupa, numerous mitochondria (M), large nuclei (N). Scale = 1 μm.

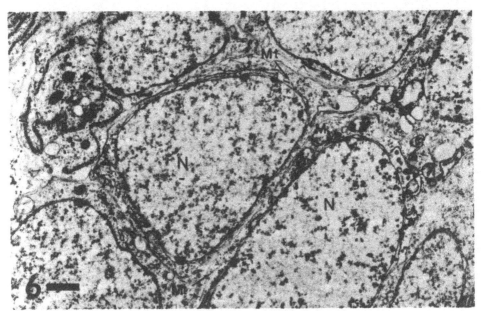

FIGURE 6. Prothoracic gland cells from 48-hr-old female pupa with large nuclei (N), few mitochondria (M) and numerous microtubules (Mt). Scale = 1 μm.

detailed ultrastructural studies of prothoracic gland cells of pupal Xyleborus is the relative abundance of microtubules in the cells of 48-hr-old as compared to 0-hr-old pupae. This finding is in close agreement with that of Birkenbeil et al. (1979) who demonstrated the presence of ecdysteroids in microtubules of prothoracic glands using immunocytochemical techniques.

The titer of ecdysteroids in the putative organ of ecdysteroid synthesis in the adult, i.e. ovary (Fig. 7 and 8), was determined by RIA on extracts of such organs dissected into insect ringers from various aged females. Ecdysteroid titers from such ovaries never equaled the whole body complement of these hormones. Ovaries of Xyleborus do not seem to be the sole organ of ecdysteroid biosynthesis, and even may simply act as a vehicle for transportation of these steroids from haemolymph into the embryos. This suggestion is further strengthened by the ultrastructural studies of ovaries at peak ecdysteroid synthesis which showed no definite ultrastructural correlates to steroid biosynthesis (Fig. 9).

Based on the above mentioned data, ecdysteroids may play roles in the adult development, especially in the ovarian events of yolk deposition and initial ovulation. This suggestion becomes more pertinent when viewed in conjunction with our findings on sorbic acid-treated (i.e., analogous to ovariectomized) beetles (Fig. 7 and 10) and on ecdysteroid contents of variously aged male beetles.

SUMMARY

The obligatory symbiosis between Xyleborus beetles and ectosymbiotic microbes involves several major lipid interdependencies. The cultivated microbes provide three types of important nutritive lipids, sterols, phospholipids and fatty acids, to the beetles. The dependency of the beetles on the ectosymbiotes for a dietary Δ^7-sterol is especially important to maintaining the symbiosis. If the female beetle is deprived of dietary Δ^7-sterol, maturation of oocytes, locomotory vigor and life span are drastically reduced. These effects involve both the ecdysteroid-hormone and the tissue-component-sterol pools in the female.

Dietary Δ^7-sterol from ectosybiotes must be transferred transovarially to progeny if normal embryogenesis is to occur. Oocyte maturation, embryogenesis, larval and pupal development, metamorphosis and adult beetle life are all dependent on specific lipid nutrition which in nature is only available from the obligate ectosymbiotes.

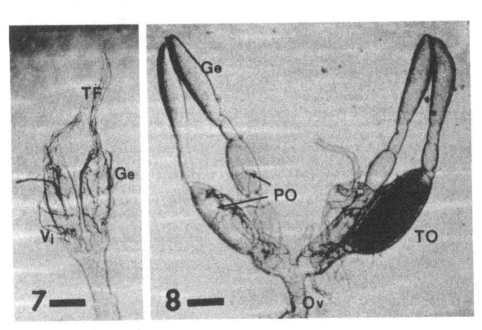

FIGURES 7 & 8.
Fig. 7. Whole mount of ovaries of X. ferrugineus, after 6 days of beetle feeding on sorbic acid-containing diet, showing a long terminal filament (TF), germarium (Ge) and vitellarium (Vi). Scale = 188 μm.
Fig. 8. Whole mount of ovaries of X. ferrugineus, after 6 days of beetle feeding on standard sawdust diet, showing one well developed terminal oocyte (To), penultimate oocyte (Po), germarium (Ge) and oviduct (Ov). Scale = 188 μm.

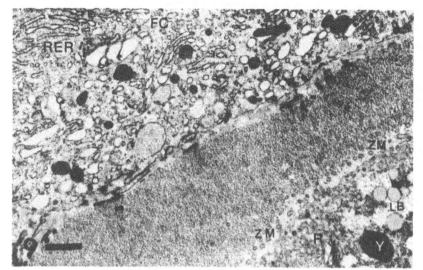

FIGURE 9. Ovary from a beetle, fed on standard sawdust
diet for 6 days. Follicle cell (FC) has numerous
rough-surfaced ER (RER). Vitellogenic oocyte is present in the ovary with variously sized yolk spheres
(Y), lipid body (LB), numerous ribosomes (R) and a zone
of microvilli (ZM). Scale = 1 μm.

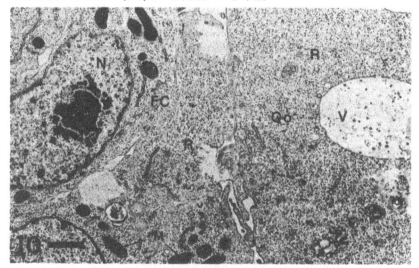

FIGURE 10. Ovary from an adult female beetle fed on sorbic acid-containing standard sawdust diet for 6 days.
There is no evidence of yolk deposition. Compact follicle cell (FC) with large nucleus (N); mitochondria
(M) and ribosomes (R) are present. In the oocyte (Oo)
there are ribosomes and a few large-sized vesicles (V).
Scale 1 = 1 μm.

ACKNOWLEDGEMENTS

Our research is supported by the College of Agricultural and Life Sciences, University of Wisconsin, Madison; and in part by research grants No. RR-00779, Division of Research Resources, and No. AG-01271, Institute on Aging, of the National Institutes of Health, and funds from the Wisconsin Department of Natural Resources to D.M.N. Figures 2, 3 and 4 are reproduced in this text with the permission of Nederlandse Entomologische Vereniging, Amsterdam, The Netherlands.

REFERENCES

Batra, L.R. (1979) Insect-Fungus Symbiosis. Nutrition, Mutualism, and Commensalism, Allanheld, Osmun & Co., Montclair, N.J.

Birkenbell, H. Eckert and M. Gersch. (1979) Electron microscopical-immunocytochemical evidence of ecdysteroids in the prothoracic gland of Galleria mellonella. Cell Tissue Res. 200, 285-290.

Bollenbacher, W. E., Vedeckis W. V., Gilbert, L. I. and O'Connor, J.D.. (1975) Ecdysone titers and prothoracic gland activity during the larval-pupal development of Manduca sexta. Develop. Biol. 44: 46-53.

Bollenbacher, W.E., Zvenko H., Kumaran, A.K. and Gilbert, L.I. (1978) Changes in the ecdysone content during the post-embryonic development of the wax moth Galleria mellonella. The role of the ovary. Gen. Comp. Endocrinol. 34: 169-179.

Borst, D. W., Bollenbacher, W. E., O'Connor, J. D., King, D.S. and Fristrom, J.W. (1974) Ecdysone levels during metamorphosis of Drosophila melanogaster. Dev. Biol. 39: 308-316.

Bridges, J.R. and Norris, D.M. (1977) Inhibition of reproduction of Xyleborus ferrugineus by ascorbic acid and related chemicals. J. Insect Physiol. 23: 497-501.

Chu, H.M., Norris, D.M. and Kok, L.T. (1970) Pupation requirement of the beetle Xyleborus ferrugineus: sterol other than cholesterol. J. Insect Physiol. 16: 1379-1389.

Chu, H.M., Norris, D.M. and Rao, K.D.P. (1980) Ultrastructure of prothoracic glands of variously aged female pupae of Xyleborus ferrugineus and associated ecdysteroid titers. Cell Tissue Res. 213: 1-8.

Chu, H.M., Norris, D.M. and Rao, K.D.P. (1982a) Sorbic acid induced differences in ultrastructural development of oocytes in ectosymbiotic female Xyleborus ferrugineus. J. Morphol., in press.

Chu, H.M., Norris, D.M. and Rao, K.D.P. (1982b) Ultrastructural evidence of accelerated aging in female

Xyleborus ferrugineus attributable to dietary sterol. *Cell Tissue Res.*, in press.

Delachambre, J., Besson, M.T., Connat, J.L. and Delbec-que, J.P. (1980) Ecdysteroid titers and integumental events during the metamorphosis of *Tenebrio molitor*. In *Progress in Ecdysone Research* (Ed. by Hoffmann, J.A.), pp. 211-234, Elsevier/North Holland Biomedical Press, Amsterdam.

Delbecque, J.P., Delachambre, J., Hirn, M. and De Reggi, M. (1978) Abdominal production of ecdysterone and pupal-adult development in *Tenebrio molitor* (Insecta, Coleoptera). *Gen. Comp. Endocrinol.* 35: 436-444.

Gande, A.R. and Morgan, E.D. (1979) Ecdysteroids in the developing eggs of the desert locust, *Schistocerca gregaria*. *J. Insect Physiol.* 25: 289-293.

Gilbert, L.I. and King, D.S. (1973) Physiology of growth and development: Endocrine aspects. *Physiology of Insecta* 1: 249-370.

Glitho, I., Delbecque, J.P. and Delachambre, J. (1979) Prothoracic gland involution related to moulting hormone levels during the metamorphosis of *Tenebrio molitor* L. *J. Insect Physiol.* 25, 187-191.

Hanaoka, K. and Ohnishi, E. (1974) Changes in ecdysone titer during pupal adult development in the silk-worm *Bombyx mori*. *J. Insect Physiol.* 20: 2375-2384.

Holman, G.M. and Meola, R.W. (1978) A high performance liquid chromatography method for the purification and analysis of insect ecdysones. Application to measurement of ecdysone titers during pupal adult development of *Heliothis zea*. *Insect Biochem.* 8: 275-278.

Hsiao, T.H. and Hsiao, C. (1975) Moulting hormone production in the isolated larval abdomen of the Colorado beetle. *Nature, Lond.* 255: 727-728.

Hsiao, T.H. and Hsiao. C. (1979) Ecdysteroids in the ovary and the eggs of the greater wax moth. *J. Insect Physiol.* 25: 45-52.

Hsiao, T.H., Hsiao, C. and de Wilde, J. (1976) Moulting hormone titer changes and their significance during development of the Colorado beetle, *Leptinotarsa decemlineata*. *J. Insect Physiol.* 22: 1252-1261.

Imboden, H., Langrein, B., Delbecque, J.P. and Luscher, M. (1978) Ecdysteroids and juvenile hormone during embryogenesis in the ovoviviparous cockroach *Nauphoeta cinerae*. *Gen. Comp. Endocrinol.* 36: 628-635.

Kok, L.T. (1979) Lipids of ambrosia fungi and the life of mutualistic beetles. In *Insect-Fungus Symbiosis. Nutrition, Mutualism, and Commensalism* (Ed. by Batra, L.R.), pp. 33-52, Allanheld, Osmun & Co., Montclair, N.J.

Kok, L.T. and Norris, D.M. (1973) Comparative sterol compositions of adult female *Xyleborus ferrugineus* and its mutualistic fungal ectosymbionts. Comp. Biochem. Physiol. 44B: 499-505.

Lagueux, M., Hirn, M. and Hoffmann, J. A. (1977) Ecdysones during ovarian development in Locusta migratoria. J. Insect Physiol. 23: 109-119.

Norris, D.M. (1972) Dependence of fertility and progeny development of Xyleborus ferrugineus upon chemicals from its symbiotes. In Insect and Mite Nutrition (Ed. by J.G. Rodriguez), pp. 299-310, Elsevier, North Holland, Amsterdam.

Norris, D.M. (1979) The mutualistic fungi of Xyleborini beetles. In Insect-Fungus Symbiosis. Nutrition, Mutualism, and Commensalism (Ed. by Batra, L.R), pp. 53-63, Allanheld, Osmun & Co., Montclair, N.J.

Norris, D.M. and Chu, H.M. (1970) Nutrition of Xyleborus ferrugineus. II. A holidic diet for the aposymbiotic insect. Ann. Entomol. Soc. Amer. 63: 1142-1145.

Norris, D.M. and Moore C.L. (1980) Lack of dietary Δ^7-sterol markedly shortens the periods of locomotory vigor, reproduction and longevity of adult female Xyleborus ferrugineus (Coleoptera, Scolytidae). Exp. Geront. 15, 359-364.

Ohtaki, T. and Takahashi, M. (1972) Induction and termination of pupal diapause in relation to the change of ecdysone titer in the fleshfly, Sarcophaga peregrina. Japan J. Med. Sci. Biol. 25: 369-376.

Rao, K.D.P., Norris, D.M. and Chu, H.M. (1981) Ecdysteroids during pupal development of female Xyleborus ferrugineus. Ent. Exp. appl. 30: 151-156.

Rao K.D.P., Norris, D.M. and Chu, H.M. (1982) Ecdysteroids in adults, ovaries and eggs of Xyleborus ferrugineus (Coleoptera, Scolytidae). J. Insect Physiol., in press.

Romer, F., Emmerich, H. and Nowock, J. (1974) Biosynthesis of ecdysones in isolated prothoracic glands and oenocytes in Tenebrio molitor in vitro. J. Insect Physiol. 20: 1975-1987.

4
Sterol Biosynthesis by Symbiotes of Aphids and Leafhoppers

H. Noda* and T. E. Mittler**

INTRODUCTION

The qualitative nutritional requirements of insects, though they may belong to different orders and families, and ingest different foods, are very uniform and similar to those of mammals. A conspicuous exception is the requirement by insects for dietary sterol. This is an indispensable nutrient for insects, because they lack the capacity for de novo biosynthesis of the steroid nucleus (Clayton, 1964). They therefore require an exogenous source of sterol to achieve their normal growth, development and reproduction (Svoboda et al., 1975). The development of insects in several orders has been compared on artificial diets with and without sterol. These studies indicate that cholesterol or one of the several phytosterols are required by the majority of insects. Sterols in insects are mainly utilized as components of cellular structures and as precursors of the moulting hormone, ecdysone. There are, however, some exceptions to the general rule that insects require dietary sterols. The demonstration that some plant-sucking homopterous insects, e.g. aphids and leafhoppers, do not require dietary sterols is an outcome of the development of chemically defined artificial diets on which these insects can be maintained in the absence of dietary sterols. On the basis of a number of nutritional studies (Clayton, 1964; Robbins et al., 1971; Svoboda et al., 1975), it is unlikely that these exceptions are due to differences in metabolism. It has been known for

* Shimane Agricultural Experiment Station, Izumo, Shimane 693, Japan.

** Department of Entomology, University of California, Berkeley, CA 94720, USA.

some time that aphids and leafhoppers harbor non-patho-
genic intracellular microorganisms within mycetocytes.
These special cells are clustered into mycetomes within
their abdomens (Uichanco, 1924; Buchner, 1965). A symbiotic, nutritional role has been attributed to these
microorganisms on the basis of the restricted diet of
these insects (Koch, 1960). Among other functions, these
symbiotes may be involved in the synthesis of sterol to
meet the host's needs.

SYMBIOTES OF APHIDS AND LEAFHOPPERS

Most species of aphids have two types of prokaryotic
endosymbiotes. One is round-oval coccoid and the other
is rod-shaped bacilliform. They are called the primary
and secondary symbiotes, respectively. The secondary
symbiotes are thought to have become associated with
aphids later than the primary ones in evolutionary history (Buchner, 1965). The pea aphid, Acyrthosiphon
pisum, and the rose aphid, Macrosiphon rosae, have both
types of symbiotes. The secondary symbiotes are contained in secondary mycetocytes situated outside the
primary mycetocytes. The cabbage aphid, Brevicoryne
brassicae, and the green peach aphid, Myzus persicae,
have only primary symbiotes. The two types of symbiote
are always restricted to their respective mycetocytes
except during infection of the embryo. Griffiths and
Beck (1973) state that the segregation of primary and
secondary symbiotes in different cells may allow the two
types to perform different functions and permit the aphid
to control these functions independently. Each of the
symbiotes is surrounded by a membranous envelope believed
to be derived from the host cell (Lamb and Hinde, 1967,
Hinde 1971a; Griffith and Beck, 1973; Iaccarino and Tremblay, 1973; McLean and Houk, 1973).
Leafhoppers also harbor intracellular symbiotes in
mycetomes in the abdominal region. Körner (1976) be-
lieves that all leafhopper symbiotes are aberrant forms
of bacteria, except for the yeast-like symbiotes. Most
leafhoppers possess two types of prokaryotic symbiotes
and some such as Euscelis plebejus, Helochara communis,
and Nephotettix cincticeps possess three (Nasu, 1965;
Chang and Musgrave, 1972; Körner, 1972, 1974; Mitsuhashi
and Kono, 1975; Schwemmler, 1974). The two common types
of prokaryotic symbiotes are referred to as a-symbiotes
and t-symbiotes. Each symbiote is surrounded by two membranes of its own as well as by a host membrane. These
microorganisms occur in two different morphological
types: vegetative and infective. The latter are transmitted transovarially to the next generation. The green
rice leafhopper, N. cincticeps, also harbors rickettsia-
like organisms throughout its body. These microorganisms

live in the nuclei as well as the cytoplasm of the host cells (Mitsuhashi and Kono, 1975). This fact raises the possibility that at least some rickettsia-like microorganisms are carried to the progeny in the sperm, whereas other leafhopper symbiotes are transmitted in the ovary.

Rice planthoppers do not have mycetomes as the other leafhoppers do. In contrast, yeast-like symbiotes are harbored in mycetocytes scattered throughout the central part of the fat body of the rice planthoppers Laodelphax striatellus, Nilaparvata lugens and Sogatella furcifera (Nasu, 1963; Noda, 1977) (Fig. 1, 2). The symbiotes are surrounded by a membranous envelope, propagate by budding, and are transmitted transovarially by a process similar to that in other leafhopper species. The small brown planthopper, L. striatellus, additionally harbors prokaryotic microorganisms that are extracellular as well as intracellular within mycetocytes (Noda and Saito, 1979a) (Fig. 3). Among plant-sucking homopterous insects, related species have similar symbiotes in corresponding sites within their bodies. This may reflect the evolutionary process of acquiring special symbiotic associations. In addition to their intracellular symbiotes, aphids and leafhoppers may harbor intestinal microorganisms (Fig. 4). Whether these microorganisms play a nutritional role, specifically one of sterol synthesis, remains to be investigated.

SYMBIOTE ELIMINATION AND ISOLATION

Although the metabolism of symbiotes and that of the insect host are undoubtedly highly integrated, the metabolic role played by the symbiotes in the insect-symbiote relationship may be elucidated by separating these two highly adapted organisms and maintaining them in isolation. This permits a comparison of the physiology of aposymbiotic insects with normal ones and a study of the biosynthetic capabilities of the microorganism in vitro.

Symbiotes can be eliminated from their hosts by several means. However, the elimination of the intracellular, obligatory symbiotes of aphids and leafhoppers (without harming the host) is rather difficult, presumably because the symbiotes do not have a stage external to the host insect even during transmission. Antibiotics incorporated into artificial diets have been used to eliminate the intracellular symbiotes of aphids and leafhoppers (Ehrhardt and Schmutterer, 1966; Mittler, 1971; Schwemmler, 1973; Griffith and Beck, 1974; Srivastava and Auclair, 1976). Removal of the symbiotes suppressed the normal growth and reproduction of these insects. Although such antibiotic-treated insects should show a requirement for dietary sterol, no improvement in the growth of larvae from antibiotic treated mother aphids

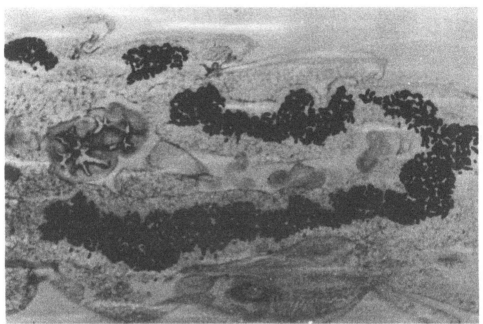

FIGURE 1. Yeastlike symbiotes; vertical section of the
abdomen of L. striatellus (x18 mag).

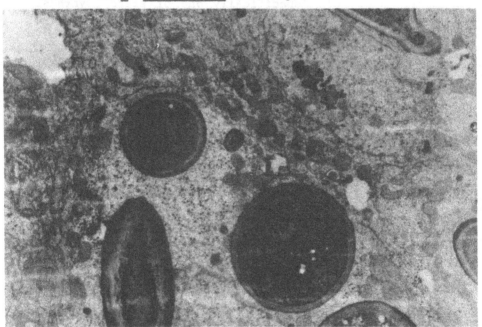

FIGURE 2. Electron micrograph of yeastlike symbiotes in
L. striatellus (8.5 mm equiv. to 1 μm).

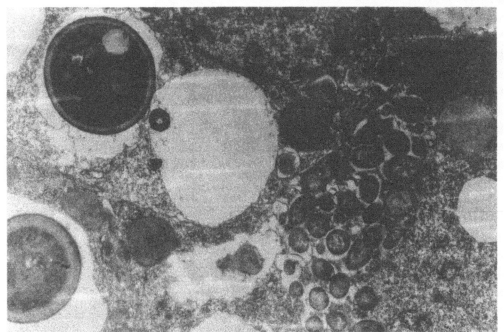

FIGURE 3. Electron micrograph of prokaryotic symbiotes in L. striatellus (10.5 mm equiv. to 1 μm).

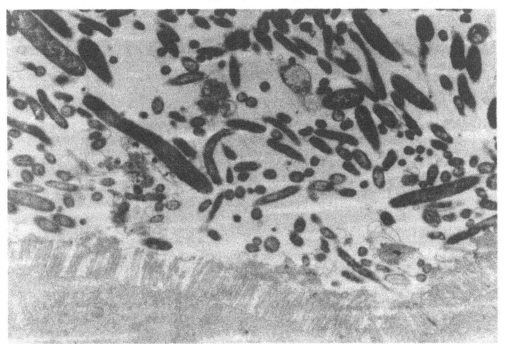

FIGURE 4. Microorganisms in the intestine of L. stria-tellus (10.5 mm equiv. to 1 μm).

was obtained by a dietary supplement of cholesterol (Mittler, 1971). This is probably due to the fact that the symbiotes are involved in other metabolic aspects of the host aphid, e.g. fatty acid, amino acid, and RNA synthesis (Houk et al. 1976; Dadd and Krieger, 1968; Mittler, 1971; Ishikawa, 1977, 1978). However, an improvement in growth of normal pea aphids was obtained when cholesterol benzoate and sulphate were added to an artificial diet (Akey and Beck, 1972). Lysozyme injection has also been used successfully with aphids and cockroaches (Ehrhardt, 1966; Daniel and Brooks, 1972), and exposure to high temperatures has been employed for the elimination of symbiotes in other insects (Richards and Brooks, 1958; Chang, 1974). In addition to any effects arising from the deprivation of symbiote-synthesized nutrients and the general destruction of the equilibrium that has evolved between the two organisms, the agents used to eliminate the symbiotes are probably directly harmful to the cellular metabolism of the host. Thus, for example, the mitochondria in a variety of cells in tetracycline treated aphids became highly abnormal (Griffith and Beck, 1974).

Culture of the symbiotes has been attempted using invertebrate tissue culture and conventional media (Orenski et al., 1965; Hinde, 1971c; Houk, 1974a; Mitsuhashi, 1975). So far, only the yeast-like symbiotes of L. striatellus and N. lugens have been maintained in permanent culture using Grace's medium (Kusumi et al., 1979, 1980; Nasu et al., 1981). Immunological methods were used for the identification of the cultured symbiotes.

STEROL BIOSYNTHESIS BY SYMBIOTES

Metabolic pathways by which three, common phytosterols (sitosterol, campesterol and stigmasterol) are converted into cholesterol by chewing insects were demonstrated for larvae of the tobacco hornworm, Manduca sexta and the silkworm, Bombyx mori (Robbins et al., 1971; Morisaki et al., 1972; Svoboda et al., 1975). For plant sap sucking insects, the metabolic conversion of phytosterol into sterols for the normal physiological function of the insect are, as yet, unknown; these processes are probably similar to those in M. sexta and B. mori, but may be dependent on, and specialized with respect to the particular phytosterols translocated in the vascular system tapped by these insects.

The finding that the green peach aphid could be maintained in permanent culture on a chemically defined artificial diet without sterol demonstrated that these aphids do not require dietary sterols (Dadd and Mittler, 1966). Subsequently, several other aphid species including the black bean aphid (Aphis fabae) and the pea aphid have been reared on such a sterol-free diet (Dadd and Krieger, 1967; Ehrhardt, 1968; Akey and Beck, 1972; Srivastava and Auclair, 1971). There is no reason to suppose that aphids have evolved a mechanism for the de novo biosynthesis of sterols, since no other insects are known to have this capability. The current hypothesis to explain the survival and development of aphids reared on sterol-free diets is that their associated symbiotes synthesize and, thereby, supply their hosts with sterols. Rearing of leafhoppers on an artificial diet free of sterols was first achieved by Mitsuhashi and Koyama (1972) with L. striatellus, using a diet modified from that for aphids. Two rice planthoppers, N. lugens and Sogatella furcifera, have since been reared on a similar sterol-free artificial diet (Koyama, 1979; Koyama and Mitsuhashi, 1980). Some other leafhoppers, however, require dietary sterol. Nephotettix cincticeps and Inazuma dorsalis were successfully reared on the artificial diets to which a cholesterol emulsion was added (Koyama, 1973). However, these species failed to grow when fed diets without such dietary cholesterol. A dietary source of sterol is also required for rearing the aster leafhopper, Macrosteles fascifrons (Hou and Brooks, 1975).

In support of the hypothesis that the symbiotes of aphids synthesize sterols, Ehrhardt (1968) reported that sodium acetate-1-^{14}C was incorporated into sterol in normal aphids (Neomyzus circumflexlus) but not in aphids that were rendered aposymbiotic by dietary chlortetracycline treatment. For A. pisum, Houk (1974b) showed that cholesterol was a constituent of the primary symbiotes; these were isolated by density gradient separation (Houk and McLean, 1974). When these symbiotes were incubated in a liquid medium with ^{14}C-acetate or ^{3}H-mevalonic acid, 2.6% and 11.8%, respectively, of the radioisotope activity of the total lipid extract was associated with the Rf region of free sterols in thin layer chromatography (Houk et al., 1976). Since a single peak with a retention time identical to cholesterol, was demonstrated in this free sterol fraction by gas-liquid chromatography, the authors concluded that the radioactivity was derived from cholesterol synthesized by the endosymbiotes of the aphid.

Griffiths and Beck (1975) proposed a mechanism by which the symbiote-synthesized cholesterol may be made available to the host. For A. pisum, small vesicles were observed in the space between the cell wall of the primary symbiotes and the membrane surrounding each symbiote. Material(s) in the vesicles appeared to enter the mycetocyte cytoplasm and thereby may be made available to the host. Hinde (1971b) had previously pointed out that a continuous lysosomal breakdown of symbiotes may be a mechanism for releasing symbiote products to the aphid.

Biosynthesis of sterols by endosymbiotes of aphids is also supported by electron microscopic histochemistry and autoradiography. Griffiths and Beck (1977a) incubated aphid tissue with digitonin (a compound which forms 1:1 molecular complexes with 3-ß-OH-sterols), and observed clearcut ultrastructural alterations at cellular sites that contain 3-ß-OH-sterols. The membranes of the primary and secondary symbiotes were significantly altered by the digitonin. Electron microscopic autoradiography with ^{3}H-mevalonic acid showed that approximately 50% of the total grains found over the mycetocytes were associated with the membranes of the primary symbiotes and the space between the membrane and the cell wall. Another 35% was associated with symbiote cytoplasm. Thus some of the label from the mevalonic acid appeared to be extensively incorporated into the symbiotes of the aphid. It should be pointed out, however, that the labelled materials in the cell walls are not necessarily sterols, since mevalonic acid can be broken down to acetate and hence metabolized to fatty acids, etc. Griffiths and Beck (1977b) also reported that many organelles in the mycetocytes were stained heavily by buffered osmium tetroxide, when the aphids were reared on a diet containing sterols. The authors concluded that the osmium deposition in the mycetocytes was due to the accumulation of precursors of sterol biosynthesis, by feedback inhibition. However, the inclusion of 3-ß-OH-sterols in the plasma membranes of prokaryotes is not, in itself, unusual. The possibility exists that these structural sterols could have been directly acquired from the host or the host's diet.

Rice planthoppers do not require dietary sterol, and the yeast-like symbiotes of these planthoppers are their probable sterol supplier. Sterol analyses by gas-liquid chromatography and gas chromatograph mass spectrometry showed that the rice planthoppers L. striatellus, N. lugens, and S. furcifera contain cholesterol, 24-methylenecholesterol and ß-sitosterol (Noda et al., 1979). 24-Methylenecholesterol is an intermediate compound in the conversion of campesterol into cholesterol in M. sexta (Svoboda et al., 1972), and it is unusual to detect such an intermediate compound in similar quantities to cholesterol. This sterol is found in the honey bee and is also a major sterol constituent of pollen (Barbier et al., 1959; Barbier and Schindler, 1959). (For further discussion of honey bee sterols, see Svoboda and Thompson in this volume). The rice plant, on which these planthoppers live, contains ß-sitosterol, campesterol and stigmasterol. In the honeydew excreted by L. striatellus, cholesterol, ß-sitosterol and a negligible amount of campesterol were detected. The cholesterol is thought to be derived from the insect whereas the ß-sitosterol is assumed to originate in the rice plant, because the planthoppers ingest a large amount of plant juice containing ß-sitosterol and excrete most of it.

The yeast-like symbiotes of L. striatellus are ther-mosensitive, and fifth instar nymphs exposed to 35°C for three days in the first instar have greatly reduced num-bers of symbiotes (Noda and Saito, 1979a). Such heat treatment was therefore used to determine whether the 24-methylenecholesterol in the insect originated in its yeast-like symbiotes. Heat-treated insects contained only a trace of 24-methylenecholesterol, and the choles-terol content was considerably below that of normal insects (Fig. 5). This finding suggested that the yeast-like symbiotes synthesize 24-methylenecholesterol. Some heat-treated nymphs failed to become adult because of an abnormality in the process of ecdysis to this stage. This phenomenon does not seem to be a direct effect of the heat treatment, because the nymphs were given the treatment for 3 days just after hatching. The deleter-ious effect of the heat treatment on ecdysis to adult-hood could partly be overcome by dietary administration of cholesterol but not by ß-sitosterol (Noda and Saito, 1979b). This suggests that cholesterol was deficient in the heat-treated planthoppers because of a reduced sup-ply of sterol from the affected symbiotes. 24-Methylene-cholesterol was not found in N. cincticeps (Noda et al., 1979). The fact that this rice leafhopper does not har-bor the yeast-like symbiotes, but possesses only the

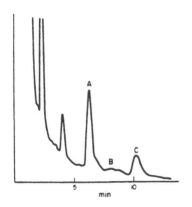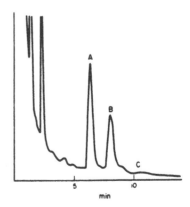

FIGURE 5. Gas chromatogram of sterols in heat-treated (left) and normal (right) fifth instar nymphs of L. striatellus. (A) cholesterol; (B) 24-methylene choles-terol; (C) ß-sitosterol. (from Noda et al. 1979).

prokaryotic symbiotes, may be the basis for this finding and explain the requirement for a dietary supply of sterol by this insect.

CONCLUDING REMARKS

Aphids and leafhoppers usually ingest the phloem sap of their host plants. Based on analyses of honeydew excreted by these insects and of plant juices (Forrest and Knights, 1972; Noda et al., 1979), this sap may contain appreciable amounts of various sterols (including cholesterol, but mostly ß-sitosterol), unless the plants are senescent (Strong, 1965). Although these insects may be able to utilize dietary sterols when available, they clearly do not depend on them entirely for their growth, development and reproduction. The fact that these homopterans have evolved the ability to harbor symbiotes and utilize the biosynthetic abilities of these microorganisms has undoubtedly enabled these insects to exploit the sieve tubes of vascular plants as an exclusive source of food. Although phloem sap is readily available to the insects by virtue of their mouthparts and the translocatory system of vascular plants, it may lack adequate levels of various essential nutrients, including specific sterols. The evidence to date suggests that such sterols are supplied to rice planthoppers by their yeast-like symbiotes and to aphids by their prokaryotic symbiotes. However, it is a curious anomaly that the latter microorganisms should be able to synthesize sterol, when in general bacterial and mycoplasmal prokaryotes lack this ability (Nes and Nes, 1980). This interesting microbiological question, as indeed the taxonomic status of the prokaryotic symbiotes and the sterol metabolism in Homoptera, in general, requires elucidation.

REFERENCES

Akey, D.H. and S.D. Beck. (1972) Nutrition of the pea aphid, Acyrthosiphon pisum: requirements for trace metals, sulphur, and cholesterol. J. Insect Physiol. 18: 1901-1914.

Barbier, M., T. Reichstein, O. Schindler and E. Lederer. (1959) Isolation of 24-methylene-cholesterol from honey bee (Apis mellifica L.). Nature 184: 732-733.

Barbier, M. and O. Schindler. (1959) Isolierung von 24-Methylene-cholesterin aus Königinnen und Arbeiterinnen der Honigbiene (Apis mellifica L.). Helv. Chim. Acta 42: 1998-2005.

Buchner, P. (1965) Endosymbiosis of animals with plant microorganisms. Interscience, New York. 901 pp.

Chang, K.P. (1974) Effects of elevated temperature on the mycetome and symbiotes of the bed bug Cimex lec-

tularius (Heteroptera). J. Invertebr. Pathol. 23:
333-340.

Chang, K.P. and A.J. Musgrave. (1972) Multiple symbiosis
in a leafhopper, Helochara communis Fitch (Cicadelli-
dae: Homoptera): envelopes, nucleoids and inclusions
of the symbiotes. J. Cell Sci. 11: 275-293.

Clayton, R.B. (1964) The utilization of sterols by insects. J. Lipid Res. 5: 3-19.

Dadd, R.H. and D.L. Krieger. (1967) Continuous rearing
of aphids of the Aphis fabae complex on sterile synthetic diet. J. Econ. Entomol. 60: 1512-1514.

Dadd, R.H. and D.L. Krieger. (1968) Dietary amino acid
requirements of the aphid, Myzus persicae. J. Insect
Physiol. 14: 741-764.

Dadd, R.H. and T.E. Mittler. (1966) Permanent culture
of an aphid on a totally synthetic diet. Experientia
22: 832-833.

Daniel, R.S. and M.A. Brooks. (1972) Intracellular bac-
teroids: electron microscopy of Periplaneta americana
injected with lysozyme. Exp. Parasitol. 31: 232-246.

Ehrhardt, P. (1966) Die Wirkung von Lysozyminjektionen
auf Aphiden und deren Symbionten. Z. Vergl. Physiol.
53: 130-141.

Ehrhardt, P. (1968) Nachweis einer durch symbiontische
Microorganismen bewirkten Sterinsynthese in künstlich
ernährten Aphiden (Homoptera, Rhynchota, Insecta).
Experientia 24: 82-83.

Ehrhardt, P. and H. Schmutterer. (1966) Die Wirkung
verschiedener Antibiotica auf Entwicklung und Symbionten künstlich ernährter Bohnenblattläuse (Aphis
fabae Scop.). Z. Morph. Ökol. Tiere 56: 1-20.

Forrest, J.M.S. and B.A. Knights. (1972) Presence of
phytosterols in the food of the aphid, Myzus persicae. J. Insect Physiol. 18: 723-728.

Griffiths, G.W. and S.D. Beck. (1973) Intracellular
symbiotes of the pea aphid, Acyrthosiphon pisum. J.
Insect Physiol. 19: 75-84.

Griffiths, G.W. and S.D. Beck. (1974) Effects of antibiotics on intracellular symbiotes in the pea aphid,
Acyrthosiphon pisum. Cell Tiss. Res. 148: 287-300.

Griffiths, G.W. and S.D. Beck. (1975) Ultrastructure of
pea aphid mycetocytes: evidence for symbiote secretion. Cell Tiss. Res. 159: 351-367.

Griffiths, G.W. and S.D. Beck. (1977a) In vivo sterol
biosynthesis by pea aphid symbiotes as determined by
digitonin and electron microscopic autoradiography.
Cell Tiss. Res. 176: 179-190.

Griffiths, G.W. and S.D. Beck. (1977b) Effect of
dietary cholesterol on the pattern of osmium deposition in the symbiote-containing cells of the pea
aphid. Cell Tiss. Res. 176: 191-203.

Hinde, R. (1971a) The fine structure of the mycetome

symbiotes of the aphids Brevicoryne brassicae, Myzus persicae, and Macrosiphum rosae. J. Insect Physiol. 17: 2035-2050.

Hinde, R. (1971b) The control of the mycetome symbiotes of the aphids Brevicoryne brassicae, Myzus persicae, and Macrosiphum rosae. J. Insect Physiol. 17: 1791-1800.

Hinde, R. 1971c. Maintenance of aphid cells and the intracellular symbiotes of aphids in vitro. J. Invertebr. Pathol. 17: 333-338.

Hou, R.F. and M.A. Brooks. (1975) Continuous rearing of the aster leafhopper, Macrosteles fascifrons, on a chemically defined diet. J. Insect Physiol. 21: 1481-1483.

Houk, E.J. (1974a) Maintenance of the primary symbiote of the pea aphid Acyrthosiphon pisum in liquid media. J. Invertebr. Pathol. 24: 24-28.

Houk, E.J. (1974b) Lipids of the primary intracellular symbiote of the pea aphid, Acyrthosiphon pisum. J. Insect Physiol. 20: 471-478.

Houk, E.J., G.W. Griffiths and S.D. Beck. (1976) Lipid metabolism in the symbiotes of the pea aphid, Acyrthosiphon pisum. Comp. Biochem. Physiol. 54B: 427-431.

Houk, E.J. and D.L. McLean. (1974) Isolation of the primary intracellular symbiote of the pea aphid, Acyrthosiphon pisum. J. Invertebr. Pathol. 23: 237-241.

Iaccarino, F.M. and E. Tremblay. (1973) Comparazione ultrastrutturale della disimbiosi di Macrosiphum rosae (L.) e Dactynotus jaceae (L.) (Homoptera, Aphididae). Boll. Lab. Ent. Agr. Portici 30: 319-329.

Ishikawa, H. (1977) RNA synthesis in aphids, Lachnus tropicalis. Biochem. Biophys. Res. Commun. 78: 1418-1423.

Ishikawa, H. (1978) Intracellular symbiont as a major source of the ribosomal RNAs in the aphid mycetocytes. Biochem. Biophys. Res. Commun. 81: 993-999.

Koch, A. (1960) Intracellular symbiosis in insects. Ann. Rev. Microbiol. 14: 121-140.

Körner, H.K. (1972) Elektronenmikroskopische Untersuchungen am embryonalen Mycetom der Kleinzikade Euscelis plebejus Fall. (Homoptera, Cicadina). I. Die Feinstruktur der a-Symbionten. Z. Parasitenk. 40: 203-226.

Körner, H.K. (1974) Elektronenmikroskopische Untersuchungen am embryonalen Mycetom der Kleinzikade Euscelis plebejus Fall. (Homoptera, Cicadina). II. Die Feinstruktur der t-Symbionten. Z. Parasitenk. 44: 149-164.

Körner, H.K. (1976) On the host-symbiont-cycle of a leafhopper (Euscelis plebejus) endosymbiosis. Experientia 32: 463-464.

Koyama, K. (1973) Rearing of Inazuma dorsalis and Nephotettix cincticeps on a synthetic diet. Jpn. J. appl. Entomol. Zool. 17: 163-166.

Koyama, K. (1979) Rearing of the brown planthopper, Nilaparvata lugens Stal (Hemiptera: Delphacidae) on a synthetic diet. Jpn. J. appl. Entomol. Zool. 23: 39-40.

Koyama, K. and J. Mitsuhashi. (1980) Rearing of the white-backed planthopper, Sogatella furcifera Horvath (Hemiptera: Delphacidae) on a synthetic diet. Jpn. J. appl. Entomol. Zool. 24: 117-119.

Kusumi, T., Y. Suwa, H. Kita and S. Nasu. (1979) Symbiotes of planthoppers: I. The isolation of intracellular symbiotes of the smaller brown planthopper Laodelphax striatellus Fallén (Hemiptera: Delphacidae). Appl. Entomol. Zool. 14: 459-463.

Kusumi, T., Y. Suwa, H. Kita and S. Nasu. (1980) Properties of intracellular symbiotes of the smaller brown planthopper Laodelphax striatellus Fallén (Hemiptera: Delphacidae). Appl. Entomol. Zool. 15: 129-134.

Lamb, K.P. and R. Hinde. (1967) Structure and development of the mycetome in the cabbage aphid, Brevicoryne brassicae. J. Invertebr. Pathol. 9: 3-11.

McLean, D.L. and E.J. Houk. (1973) Phase contrast and electron microscopy of the mycetocytes and symbiotes of the pea aphid, Acyrthosiphon pisum. J. Insect Physiol. 19: 625-633.

Mitsuhashi, J. (1975) Cultivation of intracellular yeast-like organisms in the smaller brown planthopper, Laodelphax striatellus Fallén (Hemiptera, Delphacidae). Appl. Entomol. Zool. 10: 243-245.

Mitsuhashi, J. and Y. Kono. (1975) Intracellular microorganisms in the green rice leafhopper, Nephotettix cincticeps Uhler (Hemiptera: Deltocephalidae). Appl. Entomol. Zool. 10: 1-9.

Mitsuhashi, J. and K. Koyama. (1972) Artificial rearing of the smaller brown planthopper, Laodelphax striatellus Fallén, with special reference to rearing conditions for the first instar nymphs. Jpn. J. appl. Entomol. Zool. 16: 8-17.

Mittler, T.E. (1971) Dietary amino acid requirements of the aphid Myzus persicae affected by antibiotic uptake. J. Nutr. 101: 1023-1028.

Morisaki, M., H. Ohtaka, M. Okabayashi and N. Ikekawa. (1972) Fucosterol-24,28-epoxide, as a probable intermediate in the conversion of β-sitosterol to cholesterol in the silkworm. J. chem. Soc. Chem. Commun. 1275-1276.

Nasu, S. (1963) Studies on some leafhoppers and planthoppers which transmit virus disease of rice plant

in Japan. Bull. Kyushu Agr. Expt. Stn. 8: 153-349.
Nasu, S. (1965) Electron microscopic studies on trans-
ovarial passage of rice dwarf virus. Jpn. J. appl.
Entomol. Zool. 9: 225-237.
Nasu, S., T. Kusumi, Y. Suwa and H. Kita. (1981) Sym-
biotes of planthoppers: II. Isolation of intracellular symbiotic microorganisms from the brown planthopper, Nilaparvata lugens Stål, and immunological com-
parison of the symbiotes associated with rice planthoppers (Hemiptera: Delphacidae). Appl. Entomol.
Zool. 16: 88-93.
Nes, W.R. and W.D. Nes. (1980) Lipids in evolution.
Monographs in lipid research. Plenum Press, New
York, 244 pp.
Noda, H. (1977) Histological and histochemical observation of intracellular yeastlike symbiotes in the fat
body of the smaller brown planthopper, Laodelphax
striatellus (Homoptera: Delphacidae). Appl. Entomol.
Zool. 12: 134-141.
Noda, H. and T. Saito. (1979a) Effects of high tempera-
ture on the development of Laodelphax striatellus
(Homoptera: Delphacidae) and on its intracellular
yeastlike symbiotes. Appl. Entomol. Zool. 14: 64-75.
Noda, H. and T. Saito. (1979b) The role of intracellular yeastlike symbiotes in the development of Laodelphax striatellus (Homoptera: Delphacidae). Appl.
Entomol. Zool. 14: 453-458.
Noda, H., K. Wada and T. Saito. (1979) Sterols in Lao-
delphax striatellus with special reference to the
intracellular yeastlike symbiotes as a sterol source.
J. Insect Physiol. 25: 443-447.
Orenski, S.W., J. Mitsuhashi, S.M. Ringel, J.F. Martin
and K. Maramorosch. (1965) A presumptive bacterial
symbiont from the eggs of the six-spotted leafhopper,
Macrosteles fascifrons Stal. Contr. Boyce Thompson
Inst. 23: 123-126.
Richards, A.G. and M.A. Brooks. (1958) Internal symbio-
sis in insects. Ann. Rev. Entomol. 3: 37-56.
Robbins, W. E., J. N. Kaplanis, J. A. Svoboda and M. J.
Thompson. 1971. Steroid metabolism in insects.
Ann. Rev. Entomol. 16: 53-72.
Schwemmler, W. (1973) Sprengung der Endosymbiose von
Euscelis plebejus F. und Ernährung aposymbiontischer
Tiere mit synthetischer Diät (Hemiptera, Cicadidae).
Z. Morph. Tiere 74: 297-322.
Schwemmler, W. (1974) Studies on the fine structure of
leafhopper intracellular symbionts during their
reproductive cycles (Hemiptera: Deltocephalidae).
Appl. Entomol. Zool. 9: 215-224.
Srivastava, P.N. and J.L. Auclair. (1971) An improved
chemically defined diet for the pea aphid, Acyrtho-
siphon pisum (Harris). Ann. entomol. Soc. Amer. 64:

474-478.

Srivastava, P.N. and J.L. Auclair. (1976) Effects of antibiotics on feeding and development of the pea aphid, Acyrthosiphon pisum (Harris) (Homoptera: Aphididae). Can. J. Zool. 54: 1025-1029.

Strong, F.E. (1965) Detection of lipids in the honeydew of an aphid. Nature, Lond. 205: 1242.

Svoboda, J. A., J. N. Kaplanis, W. E. Robbins and M. J. Thompson. (1975) Recent developments in insect steroid metabolism. Ann. Rev. Entomol. 20: 205-220.

Svoboda, J.A., M.J. Thompson and W.E. Robbins. (1972) 24-Methylenecholesterol: isolation and identification as an intermediate in the conversion of campesterol to cholesterol in the tobacco hornworm. Lipids 7: 156-158.

Uichanco, L. B. (1924) Studies on the embryogeny and postnatal development of the Aphididae, with special reference to the history of the "symbiotic organ", or "mycetom". Philipp. J. Sci. 24: 143-247.

5
Absorption and Utilization of Essential Fatty Acids in Lepidopterous Larvae: Metabolic Implications

Seppo Turunen*

INTRODUCTION

Since only relatively few studies have examined lipid digestion and absorption in insects, it is difficult yet to draw general conclusions on this subject, and current information is limited mainly to three orders of Insecta, namely Diptera, Orthoptera and Lepidoptera (Turunen, 1979). Data on fatty acid biosynthesis suggest that insects do not differ from other animals studied in their apparent inability to synthesize di- or polyenoic fatty acids de novo (Downer, 1978). Hence Lepidoptera, as most of the insects studied to date, have a dietary requirement for an 18-carbon di- and/or trienoic fatty acid, but the great variety of insects makes it likely that special modifications may have evolved in both the quantity and type of fatty acid necessary for a species, such as the requirement for arachidonic acid in mosquitoes (Dadd and Kleinjan, 1979).

The dietary requirement for essential fatty acid(s) is small in some species, suggesting a specific function such as hormone precursor. In this case the situation could be comparable to the vertebrate requirement of essential fatty acids as prostaglandin precursors (Gurr and James, 1975). Although phytophagous Lepidoptera may use a portion of their essential fatty acids for a comparable metabolic function, Lepidoptera characteristically have a major requirement for dietary essential fatty acid(s), linoleic ($\Delta9,12$-C18:2) or linolenic ($\Delta9,12,15$-C18:3) acid or both, and some species have adapted to diets that contain up to 50 percent linolenic acid among their fatty acids (Turunen, 1973). Such phytophagous insects digest and absorb lipids present in photosynthetic

* Department of Zoology, University of Helsinki, 00100 Helsinki 10, Finland.

tissues, and it has been suggested that these larvae might not be able to sequester enough essential fatty acids from a synthetic diet supplemented with seed oils as the main essential fatty acid source (Kastari and Turunen, 1977; Turunen, 1978).

The present article focuses on factors affecting the biological availability of essential fatty acids, based mainly on our studies of the large cabbage butterfly, Pieris brassicae. This insect is suitable for metabolic and nutritional studies because of the availability of semi-artificial diet and because of the large larval size. The species can be maintained at laboratory conditions by rearing the larvae on a diet containing linseed oil and wheat germ as sources for essential fatty acids (David and Gardiner, 1966). However, these insects may be marginally deficient in linolenic acid, as suggested by their fatty acid metabolism. The present article also contains information on possible pathways of dietary lipid absorption in relation to essential fatty acid availability. Finally some aspects of the transport of dietary lipids are considered.

In these experiments the larvae of P. brassicae were reared under controlled temperature (23°C), photoperiod (18L:6D) and relative humidity (65 per cent). The southwestern corn borer, Diatraea grandiosella, was used as an experimental insect in some studies of lipid absorption and larval hemolymph lipoproteins. The physiology and laboratory maintenance of this pyralid moth has been reviewed recently (Chippendale, 1979).

AVAILABILITY OF DIETARY LIPIDS

Pieris brassicae has a requirement for linolenic acid, which is evident, for example, as failure of pupal eclosion in insects reared on linolenate-deficient diets (Turunen, 1974b). Dietary linoleic acid has a sparing effect on the amount of linolenic acid required, and the standard artificial diet used for rearing this species contains about equal amounts of these unsaturated fatty acids. Larvae reared on this diet, in fact, accumulate more linoleic than linolenic acid into their tissue lipids. The larvae are able to regulate fatty acid absorption from the diet, as shown in one experiment by the total exclusion of erucic acid (C22:1) from tissue lipids, in spite of high dietary content of this fatty acid (Turunen, 1973). The selective absorption or exclusion of certain fatty acids may indicate either larval recognition of a particular free fatty acid or recognition of the type of lipid containing this fatty acid. A further example of this is the observed selective sequestration of linoleic and linolenic acids by larvae of the tobacco budworm, Heliothis virescens (Dikeman et al., 1981a).

We have studied some of the factors that affect the biological availability of linolenic acid. P. brassicae larvae reared on diets supplemented with seed oils as essential fatty acid sources show only limited sequestration of linolenic acid into their tissue neutral or phospholipids (Table 1). Larvae reared on two seed oil-supplemented diets containing 16.7 % and 35.7 % linolenic acid contained only 5.4% and 12.2% linolenic acid, respectively, in their tissue neutral lipids. The data also show that no erucic acid is incorporated into larval tissue lipids. Erucic acid is restricted to seed oil in Brassica rapa and insects feeding on cruciferous leaves are "normally" not exposed to this fatty acid. It is evident from these and other experiments that in larvae reared on seed oil-supplemented diets the tissue essential fatty acid content reaches a plateau which remains well blow the dietary level of these fatty acids. Why should the larvae fail to sequester more linolenic or linoleic acid from seed oils?

When P. brassicae larvae are reared on the leaves of Brassica species, the essential fatty acid content of larval tissues parallels that of the diets, exceeding 40% of fatty acids in the case of linolenic acid (Table 1). We suspect that the dietary lipid composition is important for the biological availability of the esterified

TABLE 1. Percentage fatty acid composition of diets and tissues of fifth instar Pieris brassicae larvae

	16:0	16:1	18:0	18:1	18:2	18:3	22:1	X
Diets				Total diet lipids				
I	11.0	1.1	2.1	22.0	34.0	16.7	13.1	
II	12.6	tr	3.7	15.9	32.2	35.7	-	
III	12.2	3.2	3.7	18.1	12.5	45.8	-	4.
IV	14.2	2.8	2.1	15.6	12.8	52.5	-	tr
				Tissue neutral lipids				
I	16.1	10.1	3.6	49.4	15.4	5.4	-	
II	15.4	9.3	3.3	43.8	16.0	12.2	-	
III	10.2	3.5	4.9	25.6	11.6	44.1	-	
IV	12.0	4.0	9.2	27.3	7.4	40.1	-	

Diets I and II were standard semi-artificial diets containing oil from Brassica rapa or linseed oil, respecively. Diet III was Brassica napus and diet IV B. oleracea. X = unidentified. From Turunen (1973).

fatty acids. Seed oil extracts consist chiefly of triglycerides which are apolar and not easily solubilized
in an aqueous medium without special surface-active
agents, such as the bile salts present in the mammalian
intestine. Although phytophagous insects, or insects as
a group, are thought to utilize their dietary triglycerides well (House, 1974), our studies indicate that this
is not so for P. brassicae. Comparison of the use of
dietary neutral lipids, glycolipids and phospholipids in
P. brassicae reared on Brassica oleracea has indicated
that glycolipids and phospholipids are utilized in preference to neutral lipids (Kastari and Turunen, 1977).
When lipid utiliztion was estimated from the ratio of
lipids present in the feces vs. in the diet, a low value
indicating efficient utiliztion, a ratio of 1.39 was obtained for the neutral fraction, 0.39 for the glycolipid
fraction and 0.46 for the phospholipid fraction of B.
oleracea leaves. These data indicate that the leaf-feeding larvae are adapted to absorb the more polar lipid
component of the diet. This is consistent with the data
showing that the majority of plant leaf acyl lipids consist of galactosyl diglycerides, sulfolipids and phospholipids, although neutral lipids may be significant in
some cases (Harwood, 1980; Raison, 1980).

The utilization of dietary neutral lipids is illustrated by an examination of triglyceride hydrolysis and
subsequent tissue uptake of the liberated fatty acids.
Triglycerides were found to account for about 20% of the
neutral fraction of B. oleracea leaves, but about 30% of
the neutral fraction of larval feces (Kastari and Turun-
en, 1977). Similar data were obtained for free fatty
acids which consitituted about 8% of dietary neutral
lipids but about 19% of fecal neutral lipids. These
results suggest that the larvae do not hydrolyze triglycerides as completely, for example, as mammals, which may
utilize over 90 percent of their dietary triglycerides.
Studies with radiolabeled triglyceride have given comparable results, indicating that a substantial proportion
of dietary trioleate is excreted unchanged in fifth instar P. brassicae larvae (Turunen, 1975).

A comparison of the degree of utilization of various
labeled dietary lipids is shown in Table 2. The use of
this technique in the estimation of lipid utilization is
limited by the presence of unknown lipids in the artificial diet which may affect the availability of different
lipids differently. These data are readily confirmed,
however, by quantitative thin layer chromatography of
dietary total lipids, showing that triglycerides are only
incompletely utilized by this phytophagous species. This
degree of utilization is less in actively feeding larvae
(24-48 h post ecdysis) than in larvae at the beginning
of feeding (0-15 h post ecdysis into the fifth instar).

Using thin layer chromatography and scintillation counting, fecal radioactivity in trioleate-fed larvae was
recovered from unhydrolyzed triglyceride (about 40%) and
from free oleate (about 60%). The data further show that
the overall efficiency of utiliztion of dietary free fatty acids exceeds that of triolein. It is not clear, considering the relatively high degree of free fatty acid
utiliztion, why a large proportion of fatty acids liberated from dietary triglyceride by intestinal lipase is
excreted. It is possible that hydrolysis may occur preferentially in the lower region of the intestine where

TABLE 2. Utilization of dietary lipids by fifth instar
larvae of _Pieris brassicae_

Lipid	Radioactivity (cpm/mg lipid) [a]		$\dfrac{cpm/mg\ (feces)}{cpm/mg\ (diet)} \times 100$
	Diet	Feces	
Glycerol tri-1-^{14}C oleate	10600	4080 (0-15 h) [b]	38.5
		7280 (24-48 h)	68.6
Palmitic acid-1-^{14}C	13600	2025 (24-48 h)	14.8
Linoleic acid-1-^{14}C	31500	5800 (24-48 h)	18.4
Linolenic acid-1-^{14}C	15660	1830 (24-48 h)	11.7
Phosphatidyl (N-methyl-^{14}C) choline	3150	46 (0-15 h) [c]	1.5
		89 (24-48 h)	2.8

(a) The values are averages of 2 to 3 determinations
(b) Time of feeding after the last larval ecdysis
(c) Both lipid-soluble and water-soluble radioactivity
    was determined. Radiolabeled lipids incorporated
    into the diet had the following activities: Glycerol
    trioleate, 60 mCi/mmol, palmitic acid, 50 mCi/mmol,
    linoleic acid, 50 mCi/mmol, linolenic acid, 50
    mCi/mmol, and phosphatidyl choline, 60 mCi/mmol.
    All were from the Radiochemical Centre, Amersham.

absorption is already impaired, or that the passage of food under these conditions is too rapid for more efficient hydrolysis and absorption (Turunen and Chippendale, 1977a,b). The possible selectivity in the absorption of individual fatty acids may also have some influence (Turunen, 1973). The data also indicate that a high dietary concentration of linoleic acid may have a limiting effect on the absorption of linolenic acid. Finally, it is readily shown that the availability of polyunsaturated fatty acids is markedly affected by dietary antioxidants, such as tocopherols. Tocopherols are ubiquitous antioxidants in plants and their inclusion in artificial diets at a sufficient concentration (e.g. 0.01 to 0.1%) improves the biological availability of linolenic acid in P. brassicae (Turunen, 1976).

As a result of the observed poor utiliztion of triglyceride, we turned to phospholipids as a possible source of essential fatty acids for the larvae. Radiolabeling studies confirmed previous quantitative results indicating that phosphatidyl choline is readily taken up from the diet (Table 2). Unfortunately, available phospholipids that are economically feasible as dietary constituents contain very little linolenic acid (Turunen, 1978).

ABSORPTION OF DIETARY LIPIDS

The absorption of dietary triglyceride in mammals requires the formation of small spherical particles or micelles in the intestinal lumen. These consist of the digestive products, chiefly 2-monoglycerides and free fatty acids, plus bile salts phospholipids and other minor components. In insects, complete hydrolysis of triglyceride to glycerol and free fatty avids has been observed in the isolated intestine of the locust Locusta migratoria, in which species the overall utiliztion of triglyceride was found to be efficient (Weintraub and Tietz, 1978). The monoglyceride pathway may also be of significance in glyceride absorption in the American cockroach, Periplaneta americana (Hoffman and Downer, 1979). In P. brassicae triglyceride hydrolysis yields free fatty acids and 1,2-diglycerides but it is uncertain, how significant monoglyceride formation or possible complete hydrolysis is for glyceride absorption in this species. The presence of 2-monoglycerides has not been established in the digestive or extradigestive tissues of P. brassicae or another lepidopteran, Diatraea grandiosella, and recent data have also failed to demonstrate monoglyceride in the fat body of Heliothis virescens pupae (Dikeman et al., 1981).

Some aspects of lipid digestion and absorption in insects have been reviewed recently (Turunen, 1979), and possible pathways involved in triglyceride and phospholipid digestion, absorption and their subsequent fate in P. brassicae are illustrated in Fig. 1. Hydrolysis of triglyceride to 1,2-diglyceride may be sufficient under the conditions prevailing in the midgut to allow absorption. Comparable data have been obtained also from larvae of the southwestern corn borer, Diatraea grandiosella (Turunen and Chippendale, 1977b). The absorption of partial glycerides and also of free fatty acids is followed by their incorporation into intracellular diglycerides, triglycerides and phospholipids. Triglyceride formation may represent temporary storage, the function of which is unknown. It may facilitate the maintenance of a suitable metabolite gradient between midgut cells, intestinal lumen, and hemolymph. Intracellular lipases are required for the subsequent release of diglyceride which appears to be the first metabolite to reach the hemolymph as a result of free fatty acid or triglyceride feeding (Turunen, 1975; Weintraub and Tietz, 1978; Chino and Downer, 1979).

Dietary phosphatidyl cholines or lecithins are diacyl lipids that are hydrolyzed by more than one type of

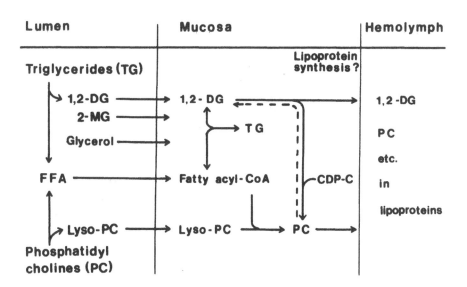

FIGURE 1. Pathways of lipid absorption, intracellular transport and release into the hemolymph in larvae of the cabbage butterfly, Pieris brassicae. DG = diglycerides, MG = monoglycerides, FFA = free fatty acids, CDP-choline = cytidine diphosphocholine.

lipase in P. brassicae. Phospholipase A activity, specific for the fatty acid in the 2-position, is responsible for the formation of the corresponding monoacyl compound, lysophosphatidyl choline in the midgut of larvae receiving choline-labeled lecithin in the diet (Fig. 1). Lysophosphatidyl cholines are surface-active compounds involved in micelle formation in mammals, and could have an analogous function in P. brassicae, facilitating glyceride absorption. Lysophosphatidyl choline itself is readily absorbed into midgut cells and resynthesized into the corresponding phosphatide by the incorporation of an activated fatty acid. It is possible, furthermore, that the triglyceride lipase could yield a lysophospholipid by hydrolyzing a fatty acid from the 1-position of lecithin. Phospholipase A activity has been reported previously in the digestive juice of three species of Lepidoptera, P. brassicae, Spodoptera exiguens and Trichoplusia ni (Somerville and Pockett, 1976). The enzyme was capable of hydrolyzing two other common plant leaf phospholipids, phosphatidyl glycerol and phosphatidyl ethanolamines.

In addition to lysolecithin, a small quantity of water-soluble radioactivity is released when phosphatidyl (N-methyl-^{14}C) choline is digested by P. brassicae (Turunen and Kastari, 1979). This suggests the presence of another type of phospholipase in the midgut lumen, specific for the phosphodiester link of lecithin. Midguts of larvae feeding on plants could contain phospholipase D which, although apparently restricted to plant tissues, might have some digestive function in the intestinal lumen of insects consuming leaves rich in this enzyme (such as cabbage, Gurr and James, 1975). Another enzyme, phospholipase C, found in animal tissues, is also capable of hydrolyzing the proximal phosphodiester link, thus yielding a water-soluble choline derivative. The choline derivative is apparently rapidly absorbed and released from the midgut into hemolymph (Turunen and Kastari, 1979).

In view of the ubiquity of phospholipids in the dietary of insects, and the polarity of these compounds, it has been hypothesized that phospholipids have an important function in glyceride transport across the mucosal cells (Turunen, 1979). In the locust, L. migratoria, data suggest that midgut phospholipids or diglycerides are precursors for hemolymph diglycerides (Weintraub and Tietz, 1978). The locusts were fed a mixture of (1-^{14}C)-trioleylglycerol and trioleyl-(2-^{3}H)-glycerol and it was found that the ^{3}H/^{14}C ratio of intestinal phospholipids and diglycerides, which were resynthesized after absorption, was similar to that of hemolymph diglycerides. These observations suggest that phospholipids absorbed from the diet, or those synthesized in midgut cells, could be important in glyceride translocation from the intestinal lumen into the hemolymph. Phospholipids, such as phosphatidyl cholines, obtained from the diet, or synthesized inside the midgut cells, could be directly released into the hemolymph, or they could be hydrolyzed and diglyceride released into the hemolymph. In the latter case the remaining activated base, e.g. CDP-choline, could be used in the phosphorylation of diglycerides (Fig. 1).

LIPID TRANSPORT IN LARVAE

The first neutral lipid appearing in the hemolymph as a result of lipid absorption seems to be diglyceride: this observation has been made in P. brassicae (Turunen, 1975), Diatraea grandiosella (Turunen and Chippendale, 1977b), Bombyx mori, Locusta migratoria (Weintraub and Tietz, 1978), and Periplaneta americana (Chino and Downer, 1979). It is also known that some larval lipoproteins contain diglycerides as principal neutral lipids, but the information on larval lipid transport remains fragmentary (Pattnaik et al., 1979; Thomas, 1979). In contrast, the involvement of diglyceride-carrying lipoproteins and female-specific vitellogenins in adult differentiation, flight and reproduction has received special emphasis (Gilbert and Chino, 1974; Chino et al., 1976; Harry et al., 1979; Gellissen and Emmerich, 1980). We have studied the lipid composition of some larval lipoproteins in the southwestern corn borer, D. grandiosella, and have found diglycerides as major neutral lipids in two electrophoretically separated lipoprotein fractions (Turunen and Chippendale, unpubl.). Data suggest that these lipoproteins carry sterols and glycerides released from midgut cells into the hemolymph of feeding larvae. The lipoprotein neutral lipids consisted of diglycerides (30 to 50%), hydrocarbons (10 to 15%), sterol esters (5 to 8%), triglycerides (10 to 25%) and free fatty acids (8 to 15%). The hemolymph of D. grandiosella also contained an electrophoretically more rapidly migrating lipoprotein fraction, with distinctly different lipid composition. The lipid composition of each fraction exhibited developmental changes. A wide spectrum of lipids in lipoproteins has been observed recently also in the cockroach, P. americana, indicating that the cockroach diglyceride-carrying lipoprotein has multiple functions in hemolymph lipid transport (Chino et al., 1982). The purified cockroach lipoprotein was also shown to be able to accept diglyceride released from midgut cells as well as from fat body. The diglycerides released from the midgut of L. migratoria have similarly been found to be incorporated into hemolymph lipoproteins (Weintraub and Tietz, 1978).

The overall pattern of lipid absorption involves hydrolysis into partial glycerides, free fatty acids, glycerol, lysophospholipids etc. These are incoporated upon absorption into midgut diglycerides, triglycerides and phospholipids. Larvae are able to regulate the absorption of at least some fatty acids, as shown by feeding the larvae erucic acid. Release of diglyceride into the hemolymph requires the presence of lipoproteins capable of accepting lipids from midgut cells. Very little is known of these larval lipoproteins and their possible functional specificity at the midgut surface.

ESSENTIAL FATTY ACID DEFICIENCY

Artificial diet-fed P. brassicae, although appearing to grow normally under laboratory conditions, synthesize substantial amounts of neutral lipids in response to an impaired availability of the essential linolenic acid. Larvae reared on the artificial diet contain on average about 22 mg total lipids per g fresh weight 52 h after the last larval ecdysis. Cabbage-fed larvae contain only about 17 mg total lipids at this stage (Kastari and Turunen, 1977). The difference in lipid was completely attributed to triglycerides which were synthesized and accumulated in the fat body. Tissue neutral lipids thus increased notably during the fifth instar, from 6.4 mg/g in newly ecdyzed larvae to 28.5 mg/g in mature, artificial diet-fed larvae. In contrast, the phospholipid contents of both types of larvae were comparable or were slightly higher in cabbage-fed larvae.
If one determines the efficiency of conversion of digested neutral lipid into insect tissue biomass (ECD) in artificial diet-fed larvae, the value obtained may exceed 250%, indicating that larvae synthesize neutral fat from dietary constituents other than lipids, viz. carbohydrates. The increased weight of artificial diet-fed larvae is not an indication of superior dietary quality: the triglyceride synthesis is a specific response of the larvae to a suboptimal level of tissue essential fatty acids. It has been shown in several studies that the deficiency of di- or polyenoic fatty acids results in an increased tissue biosynthesis of the monoenoic palmitoleic and oleic acids (Turunen, 1974a, 1976, 1978; Downer, 1978). The palmitoleic acid content is about 2 to 4% and that of oleic acid about 25 to 27% of tissue neutral lipid fatty acids in P. brassicae reared on B. oleracea, but upon experimental linolenic acid deficiency the relative proportion of palmitoleic acid increases up to 25% and that of oleic acid up to 50% of tissue neutral lipid fatty acids. Thus the increased synthesis of triglycerides can be attributed to the synthsis of monoenoic fatty acids which may have a "sparing" effect on the more

unsaturated acids. A comparable symptom of essential fatty acid deficiency is known in mammals and, for example, fish, shown as an increased synthesis of palmitoleic and oleic acid as well as triglycerides (Holman, 1971). In mammals a more specific symptom of the deficiency is an increased ratio of 20:3/20:4 fatty acids, the trienoic acid being synthesized from oleic acid. It would not be unexpected if in insects also the deficiency of 18:2 or 18:3 leads to tissue accumulation of very small amounts of long chain unsaturated fatty acids of the oleic acid family.

The accumulation of large amounts of linolenic acid, up to 45% of tissue neutral and phospholipid fatty acids, indicates that the quantitative requirement cannot be explained as a mere metabolic function for essential fatty acids in P. brassicae. Judging from the observed requirement for linolenic acid in other animal species (Tinoco et al., 1979), it is probable that the poikilothermic life of these species is one critical factor contributing to the essential role of this fatty acid. The melting point of linolenic acid is -11°C, in contrast, for example, to +10.5°C for oleic acid, and the greater fluidity of 18:3 and 18:2 is required for both lipid absorption and membrane function at low ambient temperatures.

There is a degree of sex dimorphism in the essential fatty acid requirement of P. brassicae (Turunen, 1974a). When the insects are grown on a diet providing a marginal level of polyunsaturated acids, the external symptoms of deficiency develop more readily in males than in females at pupal eclosion. Under normal dietary conditions both sexes contain about the same level of linolenic acid at the time of adult ecdysis, but the females subsequently incorporate linolenic acid into egg lipids (Turunen, 1974b). It is possible that the females have adapted to lower level of "active" linolenic acid or that the higher flight activity of males may require more linolenic acid in membrane phospholipids, such as those of flight muscle mitochondria.

When the diet of P. brassicae is supplemented with a mixture of phospholipids, the tissue content of polyunsaturated fatty acids increases and the content of monounsaturated fatty acids decreases (Turunen, 1978). The increase in tissue polyunsaturated acids is followed by decreased efficiency of conversion of food into insect tissues, as a result of fewer triglycerides synthesized by the larvae. The availability of a phospholipid rich in linolenic acid could be helpful for a systematic study of factors affecting essential fatty acid absorption in phytophagous Lepidoptera requiring linolenic acid. Such species include P. brassicae, T. ni, Autographa californica, Argyrotaenia velutinana, Estigmene acrea, Heliothis zea, and Plutella xylostella (Chippendale et al., 1964; Grau and Terriere, 1971; Terriere and Grau, 1972; Rock et al., 1965; Hou and Hsiao, 1978).

CONCLUSIONS

Some factors affecting the biological availability of essential fatty acids in the phytophagous Pieris brassicae (Lepidoptera, Pieridae), were discussed. The data suggest that availability of essential fatty acids esterified as triglycerides is rather incomplete. Seed oils used in artificial diets, furthermore, have a fatty acid composition which differs from that of plant leaves. As a result, the tissue linolenic acid content does not seem to respond to an increase in dietary seed oils beyond a certain level, but reaches a plateau which is below the dietary level of linolenic acid and significantly below the level found in the tissues of larvae reared on cabbage.

The larvae acquiring insufficient amounts of essential fatty acids synthesize "abnormally" large amounts of the monoenoic palmitoleic and oleic acids which tends to increase the accumulation of tissue triglycerides beyond what is normally found in a wild population of these larvae.

The biological availability of essential fatty acids seems to improve when the dietary fatty acids are present in phospholipids. Thus, a commercially available phospholipid rich in linolenic acid may be useful in further studies of essential fatty acid utilization and rearing methods with phytophagous, linolenic acid-requiring Lepidoptera.

Possible pathways involved in phospholipid and triglyceride hydrolysis, absorption and translocation into the hemolymph were discussed. Diglycerides released from the serosal side of midgut cells appear to be incorporated into specific larval lipoproteins serving to transport absorbed lipids to fat body and other sites of utilization.

REFERENCES

Chino, H. and R.G.H. Downer. (1979) The role of diacylglycerol in absorption of dietary glyceride in the American cockroach Periplaneta americana. L. Insect Biochem. 9: 379-382.

Chino, H., M. Yamagata and K. Takahashi. (1976) Isolation and characterization of insect vitollegenin: Its identity with hemolymph lipoprotein II. Biochim. Biophys. Acta 441: 349-353.

Chino H., H. Katase, R.G.H. Downer and K. Takahashi. Diacylglycerol-carrying-lipoprotein of hemolymph of

American cockroach: purification, characterization and function. J. Lipid Res.

Chippendale, G. M. (1979) The southwestern corn borer, Diatraea grandiosella: Case history of an invading insect. Res. Bull. Univ. Mo. Agric. Exp. Sta. no. 1031, 52 pp.

Chippendale, G. M., S. D. Beck and F.M. Strong. (1964) Methyl linolenate as an essential nutrient for the cabbage Looper, Trichoplusia ni (Hübner). Nature 204: 710-711.

Dadd, R. H. and J. E. Kleinjan. (1979) Essential fatty acid for the mosquito Culex pipiens: arachidonic acid. J. Insect Physiol. 25: 495-502.

David, W.A.L. and Gardiner, B.O.C. (1966) Rearing Pieris brassicae (L.) on semi-synthetic diets with and without cabbage. Bull. ent. Res. 56: 581-593.

Dikeman, R.N., E.N. Lambremont and R.S. Allen. (1981) Tissue specificity and sexual dimorphism of the fatty acid composition of glycerolipids from the tobacco budworm, Heliothis virescens F. Comp. Biochem. Physiol. 68B: 259-265.

Dikeman, R.N., E.N. Lambremont and R.S. Allen. (1981a) Evidence for selective absorption of polyunsaturated fatty acids during digestion in the tobacco budworm, Heliothis virescens F. J. Insect Physiol. 27: 31-33.

Downer, R.G.H. (1978) Functional role of lipids in insects. In: Biochemistry of Insects. M. Rockstein (ed.). Acad. Press, New York. 57-92.

Gellissen G. and H. Emmerich. (1980) Purification and properties of a diglyceride binding lipoprotein (PL I) of the hemolymph of adult male Locusta migratoria. J. Comp. Physiol. 136B: 1-9.

Gilbert, L. I. and H. Chino. (1974) Transport of lipids in insects. J. Lipid Res. 15: 439-456.

Grau, P. A. and J. C. Terriere. (1971) Fatty acid profile of the cabbage looper Trichoplusia ni, and the effect of diet and rearing conditions. J. Insect Physiol. 17: 1637-1649.

Gurr, M. I. and A. T. James. (1975) Lipid Biochemistry: an Introduction. Chapman and Hall, London.

Harry P., M. Pines, and S. W. Applebaum. (1979). Changes in the pattern of secretion of locust female diglyceride-carrying lipoprotein and vitellogenin by the fat body in vitro during oocyte development. Comp. Biochem. Physiol. 63B: 287-293.

Harwood, J. L. (1980) Plant acyl lipids -- Structure, distribution, and analysis. In: Biochemistry of Plants, Vol. 4. A comprehensive treatise. Lipids: Structure and function. P. K. Stumpf (ed.), 1-56. Acad. Press, New York.

Hoffman, A. G. D. and Downer, R. G. H. (1979). Synthesis of diacylglycerols by monoacylglycerol acyltransferase from crop, midgut and fat body tissues of the American cockroach, Periplaneta americana L. Insect Biochem. 9: 129-134.

Holman, R.T. (1971) Essential fatty acid deficiency. Progr. Chem. Fats Other Lipids, Vol. 9: 279-348.

Hou, R. F. N. and J.-H. Hsiao. (1978) Studies on some nutritional requirements of the diamondback moth, Plutella xylostella L. Proc. Natl. Sci. Coun. ROC, 2: 385-390.

House, H. L. (1974) Nutrition. In: The Physiology of Insecta 2nd edition, Vol. 5: 1-62. Acad. Press, London.

Kastari, T. and Turunen, S. (1977). Lipid utilization in Pieris brassicae reared on meridic and natural diets: implications for dietary improvement. Ent. exp. appl. 22: 71-80.

Pattnaik, N.M., E.C. Mundall, G.J. Trumbusti, J.H. Law and F. J. Kezdy. (1979) Isolation and characterization of a larval lipoprotein from the hemolymph of Manduca sexta. Comp. Biochem. Physiol. 63B: 469-476.

Raison, J. K. (1980) Membrane lipids--Structure and function. In: Biochemistry of Plants, Vol. 4. A comprehensive treatise. Lipid: Structure and function. P. K. Stumpf (ed.), 57-84. Acad. Press, London.

Rock, G.C., R. L. Patton and E.M. Glass. (1965) Studies on the fatty acid requirements of Argyrotaenia velutinana (Walker). J. Insect Physiol. 11: 91-101.

Somerville, M. J. and M. V. Pockett. (1976) Phospholipase activity in the gut juice of lepidopterous larvae. Insect Biochem. 6: 351-354.

Terriere, L.C. and P.A. Grau. (1972) Dietary requirements and tissue levels of fatty acids in three Noctuidae. J. Insect Physiol. 18: 633-647.

Thomas, K.K. (1979) Isolation and partial characterization of the haemolymph lipoproteins of the wax moth, Galleria mellonella. Insect Biochem. 9: 211-219.

Tinoco, J., R. Babcock, I. Mincenbergs, B. Medwadowski, P. Miljanich and M. A. Williams. (1979) Linolenic acid deficiency. Lipids 14: 166-173.

Turunen, S. (1973) Utilization of fatty acids by Pieris brassicae reared on artificial and natural diets. J. Insect Physiol. 19: 1999-2009.

Turunen, S. (1974a) Polyunsaturated fatty acids in the nutrition of Pieris brassicae (Lepidoptera). Ann. Zool. Fenn. 11: 300-303.

Turunen, S. (1974b) Lipid utilization in adult Pieris brassicae with special reference to the role of linolenic acid. J. Insect Physiol. 20: 1257-1269.

Turunen, S. (1975) Absorption and transport of dietary lipid in Pieris brassicae. J. Insect Physiol. 21: 1521-1529.

Turunen, S. (1976) Vitamin E: effect on lipid synthesis and accumulation of linolenate in *Pieris brassicae*. *Ann*. *Zool*. *Fenn*. 13: 148-152.

Turunen, S. (1978) Artificial diets and lipid balance: a quantitative study of food consumption and utilization in *Pieris brassicae* (Lepidoptera, Pieridae). *Ann*. *Ent*. *Fenn*. 44: 27-31.

Turunen, S. (1979) Digestion and absorption of lipids in insects. *Comp*. *Biochem*. *Physiol*. 63A: 455-460.

Turunen, S. and G. M. Chippendale. (1977a) Esterase and lipase activity in the midgut of *Diatraea grandiosella*: digestive functions and distribution. *Insect Biochem*. 7: 67-71.

Turunen, S. and G. M. Chippendale. (1977b) Lipid absorption and transport: sectional analysis of the larval midgut of the corn borer, *Diatraea grandiosella*. *Insect Biochem*. 7: 203-208.

Turunen, S. and T. Kastari. (1979) Digestion and absorption of lecithin in larvae of the cabbage butterfly, *Pieris brassicae*. *Comp*. *Biochem*. *Physiol*. 62A: 933-937.

Weintraub, H. and A. Tietz. (1973) Triglyceride digestion and absorption in the locust, *Locusta migratoria*. *Biochim*. *Biophys*. Acta 306: 31-41.

Weintraub, H. and A. Tietz. (1978) Lipid absorption by isolated intestinal preparations. *Insect Biochem*. 8: 267-274.

6
Metabolic Determination and Regulation of Fatty Acid Composition in Parasitic Hymenoptera and Other Animals

S. N. Thompson* and J. S. Barlow**

INTRODUCTION

The parasitic Hymenoptera are an extemely diverse group of insects comprising over 100,000 species and the largest group of parasitic animals (Askew, 1971). Their behavior and ecology have been extensively investigated as a result of the group's importance to biological control and pest management (Vinson, 1976). The parasitic habit in the Hymenoptera is generally referred to as protelean parasitism, whereby the juvenile or immature stages are parasitic on another insect or arthropod and the adult stage is "free-living." Recently, investigators have become interested in the physiology, nutrition and biochemistry of this group in order to better understand their biology in relation to animal parasitism as a whole (Vinson, 1975; Vinson and Iwantsch, 1980; Thompson, 1981). Most investigations have described the metabolites found in a number of parasite species and/or characterized the chemical alterations which occur in parasites and their hosts following parasitism. Although these studies suggest that complex physiological and biochemical interactions occur between parasitic Hymenoptera and their host species, little is known of the metabolic basis. One of the more complex and unique biochemical findings suggests that the regulation of fatty acid and lipid metabolism and the determination of fatty acid composition may be different in the parasitic Hymenoptera than in most other insects or animals in general (Barlow, 1972).

* Division of Biological Control, Univ. of Calif., Riverside, CA 92521, USA. **Department of Biological Sciences, Simon Fraser University, Burnaby, B.C., Canada.
We thank D. Jones, University of British Columbia, for his critical review.

BACKGROUND

Lipoidal compounds play important roles in the physiology and metabolism of insects and other arthropods (Downer and Mathews, 1976). The major lipid fractions of both arthropods and "higher" animals are glycerides and phospholipids. Glycerides or acylglycerols, esters of long chain fatty acids and glycerol, are the major components of depot or storage fat, and triglyceride comprises the bulk of this neutral lipid fraction with partial glyceride generally present in very small amounts. Phospholipids contain a phosphate ester moiety and phosphoglycerides, the major phospholipid fraction, are derivatives of fatty acid esters of glycerolphosphate. Phosphoglycerides, or polar lipids, are found primarily in cell membranes. Considerable information is available on glyceride and phosphoglyceride synthesis and interconversion in higher animals, and many of the component enzyme systems have been isolated and characterized (Hübscher, 1969). Although the site of de novo glyceride synthesis and storage varies considerably between species, vertebrate systems examined in depth synthesize glycerides by the following steps. After glycerolphosphate formation and fatty acid activation, glycerolphosphate acyltransferase, a particle-bound enzyme, catalyzes the synthesis of phosphatidic acid. The phosphate ester is then cleaved by phosphatidate phosphohydrolase, a soluble enzyme, yielding diglyceride. In animals, phosphoglyceride synthesis then takes place by acyltransferase activity simultaneously coupled with the replacement of the phosphate moiety through the involvement of cytidine phosphate coenzymes. During neutral glyceride synthesis the diglyceride is further esterified to yield triglyceride. In addition to the de novo synthesis from glycerolphosphate as described above, triglyceride may also result from the direct esterification of partial glyceride without the intermediate formation of phosphatidic acid. This latter acyltransferase activity is particle bound and consists of an enzyme complex comprised of monoglyceride acyltransferase and diglyceride acyltransferase as well as the fatty acid activating system and is generally referred to as glyceride synthetase (Rao and Johnson, 1966).

Less is known about the enzymatic nature of glyceride synthesis in insects. Tietz (1969) demonstrated the synthesis of glyceride through a phospholipid intermediate in a cell free preparation of locust fat body and later Peled and Tietz (1974) demonstrated the rapid acylation of parital glyceride in fat body microsomes. Studies with Ceratitis capitata demonstrated differences in glyceride synthesis between developmental stages (Municio et al., 1971; Municio et al., 1974) and more recently the synthesis of triglyceride and phospholipid in mitochondria and microsomes from the various developmental stages of that species have been examined (Megias et al., 1977; Garcia et al., 1980a).

Despite our current understanding of the bases of glyceride and phospholipid synthesis, our knowledge of the regulation of the fatty acid esterification of these lipid classes is limited. The fatty acids of most animals including insects are qualitatively similar and are comprised largely of C16 and C18 fatty acids; predominantly palmitic (C16:0) and oleic acids (C18:1Δ9) with lower levels of the corresponding monounsaturate, palmitoleic (C16:1Δ9) and saturate, stearic acid (C18:0) respectively. These are all synthesized de novo, while the polyunsaturates linoleic (C18:2Δ9,12;ω6) and linolenic acids (C18:3Δ9,12,15;ω3) are dietary requirements for many animals. Quantitatively, however, the distribution of these acids in the various lipid classes and in different tissues is markedly affected by environmental conditions, physiological factors and nutrition. Presumably all these factors play an integrated role in maintaining the fatty acid composition within adequate limits for optimal physiological function. Numerous investigations have demonstrated the physiological significance of fatty acid composition. For example, early studies with vertebrates suggested that neutral lipid saturation was an important factor in cold tolerance and acclimatization and recent studies have confirmed this (Knipprath and Mead, 1968). Studies with insects have also demonstrated a relationship between body fat saturation and environmental temperature (House et al., 1958; Keith, 1966). The importance of fatty acid composition in phospholipids has been well documented. The degree of saturation, branching and chain length of the fatty acids in phosphoglycerides are important factors in the regulation of the physical properties of liposomes and biomembranes, presumably allowing the conformational changes in protein associated with membrane function. Indeed, alterations in fatty acid saturation have been shown directly to affect membrane transport (Wilson et al., 1970) and enzyme activity (DeKruyff et al., 1973).

Nutritional studies with insects have demonstrated substantial changes in fatty acid composition in relation to dietary choline (Bridges and Watts, 1975), cholesterol (Dwivedy, 1977) and carbohydrate (Moore, 1980), and similar findings have been described for vertebrate animals (Hübscher, 1969). The fatty acid composition of the diet also affects the fatty acid composition of the body fats of animals, but the degree of influence differs widely among species as well as different tissues. Indeed, Shortland (1952) divided the animal kingdom into groups according to their ability to modify the fatty acid composition of their body fat from that of dietary fat. However, while there is little question that most animals have the ability to incorporate dietary fatty acids into their body fats and that the relative levels in the diet may be reflected to a degree in tissue lipids, studies with vertebrates as well as insects demonstrate that extreme or severe alterations in dietary fatty acid composition are by comparison little reflected in the animal's lipids.

Studies on the relationship between dietary fatty acid composition and the composition of insect body fats are numerous, but most simply involve a comparison of the animal's composition when reared on various crude diets. Extreme alterations in the fatty acid content of Pablum-based diets brought about by supplementation with stearic, palmitoleic and linoleic acids in the free acid as well as triglyceride form, resulted in only minor corresponding alterations in the body fats of <u>Galleria</u> <u>mellonella</u> (Thompson and Barlow, 1972a). In contrast, <u>Schaefer</u> (1968) reported differences in the fatty acid composition of <u>Heliothis</u> <u>zea</u> reared on artificial media containing wheat germ compared with a diet of broad bean leaves; and Vanderzant (1968) reported similar findings when <u>H</u>. <u>zea</u> was reared on artificial diets containing corn oil compared with linseed oil. Barnett and Berger (1970) demonstrated alterations in the fats of <u>H</u>. <u>zea</u> reared on wheat germ diets containing either safflower oil or coconut oil. The alterations in fatty acid composition in these studies with <u>H</u>. <u>zea</u>, however, were primarily in the content of polyunsaturates. For example, in <u>H</u>. <u>zea</u> reared on broad bean leaves, 45% of its fatty acids were polyunsaturated compared with 3% when fed the artificial media with wheat germ. The situation with the polyunsaturated fatty acids is unique since these are not synthesized <u>de</u> <u>novo</u> or are not synthesized in sufficient quantity to contribute significantly to body fat deposition (see Chapter by Dadd for latest information). Thus, they are only present in the animal's fats when present in the diet.

Unfortunately, only a few studies on insects have been made with chemically-defined media under rigid environmental control and attempts have not been made to correlate dietary effects with metabolic activity. Studies with chemically-defined diets have been made with three species of Diptera; <u>Musca</u> <u>domestica</u> (Barlow, 1966a), <u>Lucilia</u> <u>sericata</u> (Barlow, 1966b) and <u>Agria</u> <u>housei</u> (Barlow, 1965). Although the effects were not identical in all three species, considerable variation in dietary fatty acid composition had limited effects on body fats except in the case of polyunsaturated fatty acids. Indeed, the composition was similar even when the insects were reared on fatty acid-free media. Most dipterous

species have a characteristically high level of palmitoleic acid (C16:1) which is, in general, a minor component
of most animal fats (Thompson, 1973). Its presence then
serves as a marker to more critically indicate the
effects of dietary composition on that of body fats.
Regardless of alteration in dietary level, including de-
letion, palmitoleic acid was retained at high levels in
all three species. Furthermore, although polyunsaturates
were present only when provided in the media, their levels were not directly related to their relative amounts
in the diet. Studies with other insect groups which display unique compositions also support the view that the
dietary fatty acid composition is severely moderated by
metabolic processes. For example, various homopterous
species display unusually high levels of myristic (C14:0)
acid, a trace component in most animal fats, and these
levels are maintained regardless of dietary fatty acid
composition (Strong, 1963).

Although most studies with insects have not attempted
to correlate the effects of dietary fatty acid composition to metabolic activity, any correlations must, indeed, be considered the result of biochemical and physiological regulation. Much of the variation in body fat
composition may be due to the physical significance of
saturation. For example, despite alteration in the acyl
specific composition of body fat of H. zea in response
to diet, the total unsaturated fatty acids always comprised 68 to 70% of the total (Vanderzant, 1970). Similar results were reported by Schaefer (1968) and Barnett
and Berger (1970).

One of the most detailed analyses of the effects of
dietary fatty acid composition on tissue fatty acid composition was that made by Caster et al. (1966) with rats.
The effects of 21 mixtures of 12 fatty acids fed at the
10% caloric intake level on liver fatty acid composition
were examined. Metabolic interactions and conversions
and their nutritional relationships were then determined
and equations obtained for estimating the tissue levels
of seven fatty acids from the fatty acids levels in the
diet and the dietary levels of four fatty acids from the
fatty acids levels in the tissue all expressed as percentages of the total fatty acids present. For example,
the tissue C18:1 (oleic acid) level was strongly correlated with a complex relationship of metabolic and
dietary factors defined by the equation: C18:1 (tissue) =
32.02 + 1.22 C14:0 (diet) + 0.54 C16:0 (diet) + 0.59
C18:0 (diet) - 2.26 C18:2 (diet) - 1.69 C18:3 (diet).
The complex involvement of both metabolic and dietary
factors is evident. Of special interest were the high
positive diet-tissue correlations between acids of the
linoleic acid (ω6) and linolenic acid (ω3) families,
although these were shown to be affected by the levels

of certain saturated and monosaturated acids as well. The authors concluded that increases in dietary levels of C14:0, C16:1, C18:2, C18:3 and C20:1 resulted in increased tissue levels but negative correlations or no correlations were obtained for C16:0, C18:0 and C18:1.

In vitro investigations have demonstrated several metabolic steps that may regulate the composition of fatty acids in complex lipid classes. Acyl substrates for glyceride synthesis are provided as CoA esters produced largely by ATP dependent long chain acyl-CoA synthetases. The acyl components are presumably derived from the free fatty acid pool, a very minor lipid component in cells, arising by lipolytic hydrolysis and de novo fatty acid synthesis. De novo fatty acid synthesis occurs by the repetitive condensation of malonyl-CoA, catalyzed by a soluble, cytoplasmic multi-enzyme complex generally referred to as fatty acid sythetase, and represents the first step which may influence the composition of the free fatty acid pool and thereby affect the final composition of glyceride fatty acids. Following the synthesis of long chain acyl-CoA by synthetase in animals, the ester is hydrolyzed by a thioesterase and the final product, free fatty acid, is released. In mammalian systems this product is primarily palmitic acid (C16:0), although the pattern of fatty acids produced by synthetase activity has been shown to be affected by the concentration of malonyl-CoA (Tame and Dils, 1969). By varying the activity of acetyl-CoA carboxylase and thus malonyl-CoA concentration, the chain length of fatty acids synthesized by guinea pig intestinal mucosa synthetase was markedly affected and substantial amounts of myristic and lauric acids, in some cases over 50% of the total, were produced at decreasing malonyl-CoA levels. In studies with insects, crude sythetase preparations from Lucilia sericata incorporated malonyl-CoA into stearic and palmitic acids in a consistent 1.5:1 ratio throughout several purification steps (Thompson et al., 1975), and homogenates of Aldrichina grahami synthesized myristic and palmitic acids de novo (Takaya and Miura, 1968). Studies with highly purified synthetase preparations demonstrated palmitic acid as the single product in Ceratitis capitata (Municio et al., 1977), but in the aphid, Acyrthosiphon pisum, myristic acid was the major product (G. Bloomquist, personal communication).

Two additional fatty acid synthesizing enzyme systems which contribute substrate to the free fatty acid pool for glyceride synthesis are the elongation systems and oxidative desaturation. Elongation may be by a mitochondrial system catalyzing a modified reversal of ß oxidation using acetyl-CoA or by a microsomal system utilizing malonyl-CoA to elongate pre-existing long chain acetyl-CoA, primarily palmityl-CoA and yielding stearic acid (Pynadath, 1969). Numerous studies have indicated the presence of such elongation systems in insects. Although some in vitro investigations have been made (Municio et al., 1972; Gonzales-Buitrago, et al., 1979) most studies have been in vivo and demonstrate that the pattern of radioactivity incorporated from labelled acetate or other substrates and the distribution of activity in fatty acids synthesized is consistent with the elongation system (Thompson and Barlow, 1971; Municio et al., 1974). Fatty acid desaturases are microsomal-bound oxidases catalyzing the desaturation of myristic, palmitic and stearic acids to their corresponding 9 monounsaturates. Howling et al., (1972) have demonstrated that the desaturases from several vertebrate systems exhibit fatty acid substrate specificity. Both in vivo and in vitro studies have established the presence of desaturase activity in insects (Barlow, 1966b; Tietz and Stein, 1969; Municio et al., 1972, 1974; Gonzales-Buitrago et al., 1979). The regulation of the activities of the above fatty acid synthesizing systems (all of which are active in most animals) plays an important role in determining the composition of fatty acids available for glyceride synthesis.

Fatty acid activation prior to glyceride synthesis is catalyzed by acyl-CoA synthetase. Numerous studies with a variety of vertebrate systems have demonstrated substrate specificity, but little consistency has been observed between systems and it is suggested that several enzymes are involved (Groot, et al., 1976). Recent studies with brown adipose tissue demonstrated that among C16 and C18 fatty acids, palmitic acid was the preferred saturated substrate and oleic and linoleic acids the preferred unsaturated substrates (Normann and Flatmark, 1980). The chain length specificity observed compared well with the fatty acid composition of that tissue. Municio et al. (1975a) examined acyl-CoA synthetase activity in larval homogenates of the insect Ceratitis capitata and found that linoleic acid was incorporated to a greater extent than oleic acid.

The formation of phosphatidic acid from glycerol-phosphate and acyl-CoA during the first stages of glyceride synthesis is catalyzed by glycerolphosphate acyltransferase and C16 and 18 fatty acids are the preferred substrates in a number of vertebrate systems (Kuhn, 1967). Lysophosphatidic acid is an intermediate in many cases and Lands and Hart (1965) demonstrated that two separate enzymatic acylations actually occur. Furthermore, these authors demonstrated that the rate of acyl-CoA incorporation in microsomes of rat liver was substrate specific; C18:2 > C18:1 > C14:0 >C16:0 > C18:0. Studies utilizing enzymatic acyl-CoA generating systems to provide the acyl-CoA substrate demonstrated distinct positional specificity (Possmayer et al., 1969). This assay procedure supplies acyl-CoA substrate in the free monomeric active form and overcomes the inhibitory effect of acyl-CoA supplied exogenously in micellar form. In that study saturated fatty acids were incorporated primarily into position 1 and unsaturated in position 2.

Although phosphatidate phosphohydrolase does not directly affect the acyl component of glyceride intermediates, studies by McCaman et al. (1965) indicate that the enzyme's activity is affected by the fatty acid composition of the phosphatidic acid substrate and it would, therefore, be expected to indirectly affect the fatty acid composition of the diglyceride product released.

Following the hydrolysis of the phosphate ester of phosphatidic acid and the release of 1,2 diglyceride, microsomal-bound diglyceride acyltransferase catalyzes further acylation and synthesis of triglyceride. The fatty acid composition of the diglyceride substrate has been demonstrated to be important in several vertebrate systems. For example, Ailhaud et al. (1964) found that 1,2 dipalmitin was a poor substrate whereas 1,2 diolein was acylated at the highest rates and mixed diglycerides were intermediate. With regard to acyl-CoA specificity, stearyl-CoA was incorporated at greater rates than oleyl-CoA.

Preformed partial glyceride is used in the synthesis of triglyceride as a substrate for diglyceride acyltransferase and monoglyceride acyltransferase. The latter enzyme from several vertebrate intestinal mucosa preparations catalyzes the acylation of DL1 and 2 monoglycerides to diglyceride but the enzyme has a higher affinity for the 2 isomer (Ailhaud et al., 1964). The fatty acid component of the monoglyceride is important as Clark and Hübscher (1963) reported that monoolein was utilized at greater rates than monostearin and monolinolein. The enzyme also exhibits acyl-CoA substrate specificity. Utilizing an acyl-CoA generating system, Brindley and Hübscher (1966) demonstrated preferential uptake of C16:0 and C18:1 as compared with C18:0 into both 1 and 2 monopalmitin. Ailhaud et al. (1964) found that the fatty acid is added preferentially to the outer primary hydroxyl group of the monoglyceride and the degree of preference was influenced by the fatty acid composition of both the monoglyceride and acyl-CoA substrates. With regard to triglyceride synthesis from diglyceride, Johnson et al. (1967) demonstrated that diglyceride produced from preformed monoglyceride by monoglyceride acyltransferase did not equilibrate with that formed de novo from glycerolphosphate suggesting that two separate diglyceride acyltransferase enzymes are involved. Such independent activity of acyltransferase enzymes supports the view that two totally separate pathways of triglyceride synthesis occur and these have been termed the glycerol-phosphate pathway, referred to as de novo triglyceride synthesis in the current report, and the monoglyceride pathway which is carried out by the activity of glyceride synthetase.

Less is known concerning the substrate specificity of the various acyltransferase enzyme systems involved in glyceride synthesis in insects. Municio et al. (1971) demonstrated that stearic acid was incorporated into glycerides with far less metabolic efficiency than palmitic acid by microsomes of larval homogenates of Ceratitis capitata and suggested that this specificity was responsible for the low level of stearic acid in the body fats of that species. Similar results with the acyl-CoA derivatives of palmitic and stearic acids were obtained by Barlow et al. (1980) with Galleria mellonella. Later Municio et al. (1975a) reported that larval C. capitata microsomes incorporated palmitic and linoleic acid into triglyceride more efficiently than oleic acid. Municio et al. (1975b) demonstrated the positional specificity of fatty acid incorporation into triglyceride of C. capitata larval microsomes and found that positions 1 and 3 were composed primarily of saturated fatty acids and position 2 of unsaturated fatty acids. More specifically, however, palmitic acid was incorporated equally into positions 1 and 3, myristic acid into position 3, linoleic acid into position 2 and oleic acid distributed between position 2 and 3 with somewhat greater incorporation into position 3. Kinetic analysis of fatty acid incorporation into triglyceride, demonstrated an initial rapid incorporation at position 3 proposed to result from acylation of endogenous diglyceride. Lower initial incorporation at positions 1 and 2 suggested fatty acid incorporation by de novo diglyceride synthesis followed by an acyl-exchange mechanism that resulted in a specific non-random fatty acid distribution in the triglycerides. Indeed, the kinetics observed were consistent with the stereospecific fatty acid composition of the triglyceride fraction. Barlow et al. (1980), examining stoichiometric relationships of acyl-CoA incorporation into glycerides, found the amount of acylation required for observed glycerophosphate incorporation was much greater than could be accounted for by the acyl-CoA supplied. They suggested that exogenous acyl-CoA positively effects the release of endogenous acyl groups which then also serve as substrate for acylation. It is evident that endogenous substrates in such in vitro studies must be considered before adequate determinations can be made of the substrate specificity of acyl-transferase enzymes. Barlow and Jones (1980) reported that the addition of bovine serum albumin reduced the ability of G. mellonella microsomes to acylate glycerolphosphate with endogenous acyl substrate and recently, studies have been made to determine the substrate specific nature of the acyltransferase of that insect and a second species, L. sericata.

The results of experiments utilizing an acyl-CoA generating system and showing the rates of the acylation of glycerolphosphate with myristic, palmitic, palmitoleic, oleic and linoleic acids by microsomes of G. mellonella and L. sericata are shown in Figure 1. Palmitic acid was the preferred substrate in G. mellonella and is the major fatty acid in the glycerides of the species, while the rapid incorporation of palmitoleic acid in L. sericata relative to G. mellonella is consistent with the high characteristic level of that acid in the lipids of L. sericata. However, direct correlation between the

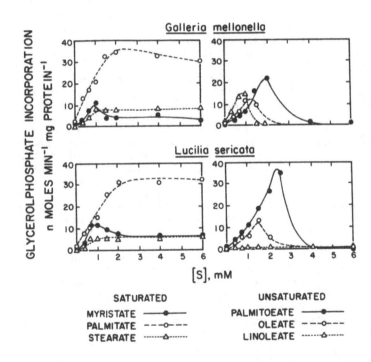

Figure 1. In vitro incorporation of glycerolphosphate into the microsomal lipids of Galleria mellonella and Lucilia sericata and the effects of fatty acid substrate concentration (D. Jones and J.S. Barlow, unpublished).

TABLE 1
Substrate specificity of transacylation in Galleria mellonella and Lucilia sericata (D. Jones and J. S. Barlow, unpublished).[1]

| Fatty acid | Lucilia sericata | | | | Galleria mellonella | | | |
| | Relative maximum velocity[2] Duplicate | | % composition[3] | Relative abundance[4] | Relative maximum velocity[2] Duplicate | | % composition[3] | Relative abundance[4] |
	I[5]	II			I[5]	II		
C14:0	2.1	2.1	2	0.6	1.3	2.4	1	0.3
C16:0	5.7	3.2	29	11.0	4.8	5.3	48	30.0
C16:1	6.5	4.9	20	7.8	2.8	2.1	4	1.0
C18:0	1.0	1.0	3	1.0	1.0	1.0	2	1.0
C18:1	2.1	1.4	39	15.0	1.6	---	40	25.0
C18:2	0.1	0.2	7	2.8	2.1	1.7	6	3.6

[1]It should be noted that both acyl-CoA synthetase and acyltransferase activities are involved in the reaction since free fatty acid served as substrate.

[2]Maximum velocity of each fatty acid at optimal substrate concentration/maximum velocity with stearic acid at optimal substrate concentration.

[3]Based on a composition of these acids alone in total body fat and not glyceride.

[4]Actual percent composition of each individual fatty acid/actual percent composition of stearic acid (from Thompson and Barlow, 1972b).

[5]Data determined from Figure 1.

rates of incorporation of these and the other fatty acids relative to their specific levels in the glycerides were not straightforward. These relationships were examined more precisely by comparing the estimated maximum velocity of the incorporation of the various fatty acids to their distribution and relative abundance in body fat glycerides (Table 1). This study and those cited above point out the multicomponent nature and complexity involved in the metabolic determination of fatty acid composition of animal body fats.

With regard to the incorporation of exogenous partial glycerides, Tietz et al. (1975) demonstrated the specificity of monoacylglycerol acyltransferase activity in locust fat body microsomes for 2-monoacylglycerol. In Periplaneta americana, Hoffman and Downer (1979) also demonstrated the acylation of partial glyceride, with 2-monoacylglycerol the preferred substrate. Garcia and Municio (1980) reported that the incorporation of oleic acid into triglycerides of C. capitata larval microsomes was consistent with synthesis of triglyceride by the esterification of pre-existing partial glyceride and most recently Garcia et al. (1980a,b) characterized the synthesis of triglyceride by that pathway.

STUDIES WITH PARASITIC HYMENOPTERA

In contrast to the maintenance of the fatty acid composition of most animals within rather specific limits, Bracken and Barlow (1967) reported that the fatty acid composition of the total body fats in the ichneumonid parasite, Exeristes comstockii, was nearly identical to that of its host species. Those workers reared the polyphagous species on three insect hosts, G. mellonella, L. sericata and Neodiprion sertifer, which differed considerably in fatty acid composition. In each case, the fatty acid composition of the parasite was indistinguishable from the host on which it was reared. Furthermore, the compositions of the parasites were retained through pupation and eclosion and did not revert to a pattern which might be considered typical of that species. Thompson and Barlow (1970) reported the same occurrence in the closely related species, Itoplectis conquisitor, when reared on G. mellonella and Ostrinia nubilalis (Fig. 2). Later, these workers surveyed 30 species of parasites from 5 families comparing the parasites' fatty acid composition to those of their host species (Thompson and Barlow, 1974). Since alternate hosts species with different fatty acid composition could not be used in most cases, quantitative estimates of the similarity between the individual parasite and host compositions were made by determining two statistical indices of similarity. The first, referred to as the mean square distance (MSD), was the variance of the quantitative differences between the individual fatty acids of each parasite and host; the second, the variance of the differences from unity of the ratios of the individual fatty acids, was developed specifically for those analyses and termed the mean square distance of the ratios (MSDR). The MSDR more accurately reflects similarity since it is dependent upon proportionality rather than absolute differences and, therefore, reduces the effect of a proportionately small difference in abundant fatty acids and increases the effect of a difference which is proportionately large in less abundant fatty acids. Of the parasitic Hymenoptera examined, those of the family Ichneumonidae exhibited fatty acid compositions most similar to their host species' as indicated by the low

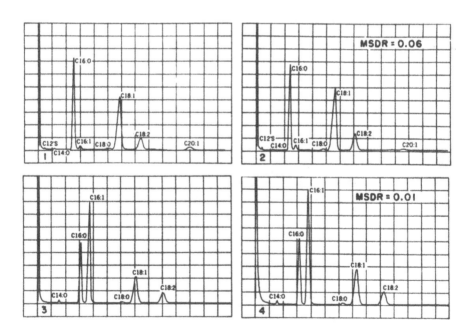

FIGURE 2. The fatty acid "patterns" of Galleria mellonella (1), Itoplectis conquisitor reared on G. mellonella (2), Ostrinia nubilalis (3), I. conquisitor reared on G. nubilalis (4) and corresponding similarity indices (after Thompson and Barlow, 1970).

TABLE 2
The similarity of the fatty acid compositions of parasitic Hymenoptera and their hosts (after Thompson and Barlow, 1974).

Parasite and Host	MSD	MSDR
Superfamily Ichneumonoidea		
Family Aphidiidae		
1. Aphidius smithi	384.58	4.29
Acyrthosiphon pisum (Hom.)		
Family Braconidae		
1. Agathis gibosa	9.86	0.13
Gnorimoschema operculella (Lep.)		
2. Agathis unicolorata	65.70	1.66
G. operculella (Lep.)		
3. Apanteles scutellaris	80.48	3.18
G. operculella (Lep.)		
4. Apanteles subandinus	74.49	2.28
G. operculella (Lep.)		
5. Aphaereta pallipes	135.72	0.71
Agria housei (Dip.)		
6. Bracon hebetor	12.85	0.77
Paramyelois transitella (Lep.)		
7. Chelonus blackburni	107.00	5.20
Sititroga cerealea (Lep.)		
8. Chelonus texanus	53.19	1.14
G. operculella (Lep.)		
9. Macrocentrus ancylivorus Rohwer	39.14	2.60
G. operculella (Lep.)		
10. Microplitis croceipes	102.96	0.75
Heliothis zea (Lep.)		
11. Microplitis plutellae	78.86	0.26
Plutella xylostella (Lep.)		
12. Peristenus stygicus	7.15	0.12
Lygus hesperis (Hom.)		
Family Ichneumonidae		
1. Campoplex haywardi	130.12	4.16
G. operculella (Lep.)		
2. Diadegma insularis	20.99	0.05
P. xylostella (Lep.)		
3. Exeristes comstockii	20.38	0.04
Lucilia sericata (Dip.)		
4. Exeristes roborator	27.00	0.03
Pectinophora gossypiella (Lep.)		
5. Hyposoter exigua	37.84	0.24
Trichoplusia ni (Lep.)		
6. Itoplectis conquisitor	3.86	0.01
Ostrinia nubilalis (Lep.)		
7. Pimpla turionellae Linneaus	21.27	0.03
Galleria mellonella (Lep.)		
8. Pleolophus basizonus	1.45	0.01
Neodiprion sertifer (Hym.)		

TABLE 2 (cont'd)

Parasite and Host	MSD	MSDR
9. Pristomerus hawaiiensis G. operculella (Lep.)	105.06	3.59
10. Temelucha sp. G. operculella (Lep.)	29.13	0.90
Superfamily Chalcidoidea		
Family Pteromalidae		
1. Muscidifurax raptor M. domestica (Dip.)	29.95	0.29
2. Nasonia vitripennis M. domestica (Dip.)	67.93	3.54
3. Spalangia cameroni Musca domestica Linneaus (Dip.)	10.47	0.04
4. Spalangia endius M. domestica (Dip.)	20.58	0.16
Family Eulophidae		
1. Dahlbominus fuscipennis Neodiprion lecontei (Hym.)	24.55	0.14
2. Encarsia formosa Trialeurodes vaporarum (Hom.)	117.81	1.01

indices (Table 2). However, this similarity was not restricted to or universal within that family.

The initial studies above coincided with others on parasitic helminths which demonstrated that certain species had fatty acid compositions qualitatively similar to their hosts. Later studies with helminths demonstrated the inability of several parasitic forms to synthesize saturated fatty acids, desaturate dietary fatty acids or ß oxidize fatty acids (Meyer and Meyer, 1972). Fatty acid for glyceride synthesis is, therefore, provided by the diet and control over glyceride composition presumably results from selective acylation (Fairbairn, 1970). In contrast to the situation with parasitic helminths, hymenopterous parasites demonstrated a marked capacity for fatty acid synthesis.

In vivo studies with I. conquisitor, E. comstocki and Exeristes roborator, three parasite species which "duplicate" the fatty acid composition of their hosts, have demonstrated that these species incorporate a variety of radiolabelled precursors into saturated and monounsaturated fatty acids. Studies on the incorporation of ^{14}C-1-acetate in E. roborator reared on G. mellonella demonstrated an immediate and rapid synthesis of fatty acids (Fig. 3). The specific activity increased sharply with time and equilibrated after approximately 30 minutes. Significant amounts of label were incorporated into palmitic and stearic acids although small amounts were also found in myristic, palmitoleic and oleic acids. The relative specific activity (% dpm/% composition) of the individual C16 and 18 fatty acids showed no significant change over time for any one acid, that is, the relative rates of incorporation into the individual fatty acids did not change with respect to one another (Table 3). Decarboxylation of palmitic and stearic acids indicated their synthesis by recognized synthetic pathways. Specifically, considering the proportion of label expected in terminal carbon resulting from de novo synthesis as 12.5% (alternate carbons labelled) and 11.1% in palmitic and stearic acids, respectively, and from elongation as 100% in both cases, the data indicate that the distribution of label in palmitic acid is consistent with de novo synthesis and the distribution in stearic acid, 75% uniformly labelled and 25% terminally labelled, is consistent with the elongation of de novo synthesized palmitic acid (Table 4). Similar results were reported by Thompson and Barlow (1972b) for E. comstockii reared on G. mellonella following 24 hours of incubation with ^{14}C-1-acetate. In this case, however, significant radioactivity was also incorporated into monounsaturated fatty acids. This was not the case in the studies cited above on E. roborator carried out at shorter incubation times. Decarboxylation, however, again demonstrated activity distributions consistent with de novo synthesis of palmitic as well as myristic acid and the elongation of palmitic to stearic acid. The distribution of activity in the monounsaturates in E. comstockii was consistent with the desaturation of palmitic and stearic acids.

Experiments with the parasite I. conquisitor reared on G. mellonella demonstrated the synthesis of fatty acids and distribution of label among the fatty acids to be similar to that found in E. comstockii reared on that host (Barlow and Bracken, 1971).

The synthesis of fatty acids in E. comstockii was also compared with that in two hosts, G. mellonella and L. sericata (Thompson and Barlow, 1972b). The specific activities of fatty acids in the parasite were significantly higher than in either host, indicating a much greater rate of synthesis and turnover. Also of interest in that study was the finding that although the level

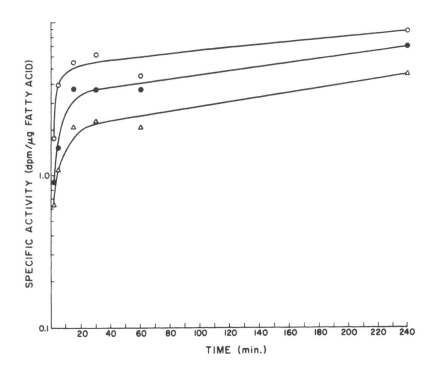

FIGURE 3. Initial in vivo incorporation of radioactivity from ^{14}C-1-acetate into total fatty acids (▲), C18:0 (O) and C16:0 (●) of Exeristes roborator (S.N. Thompson and J.S. Barlow, unpublished).

TABLE 3
Composition and relative specific activities of fatty acids synthesized in vivo by *Exeristes roborator* from ^{14}C-1-acetate (S. N. Thompson and J. S. Barlow, unpublished).

Relative Specific Activity

Fatty Acid	Percent composition[1] ($\bar{x} \pm$ S.D.)	% dpm % Composition 2 min.	5 min.	15 min.	30 min.	60 min.	4 hr.	$\bar{x} \pm$ S.D.
C16:0	28.36 ± 7.71	1.87	1.73	2.26	2.15	2.36	2.42	2.17 ± 0.27
C16:1	4.30 ± 1.17	2.14	0.82	1.34	2.34	2.19	2.01	1.85 ± 0.57
C18:0	3.24 ± 0.59	3.68	4.44	3.47	3.73	2.99	3.10	3.51 ± 0.72
C18:1	49.06 ± 7.12	0.27	0.17	0.16	0.18	0.18	0.25	0.19 ± 0.05

[1] C18:2, C18:3 and C20:1 comprised the remaining fatty acids.

TABLE 4.
Theoretical vs. observed radioactivity recoveries from decarboxylation of C16:0 and C18:0 synthesized by Exeristes roborator from [14]C-
1-acetate (S. N. Thompson and J. S. Barlow, unpublished).

Fatty Acid	Expected recovery c̄ uniform labelling (de novo synthesis)[1]	Expected recovery c̄ terminal labelling (elongation)	Average recovery observed
C16:0	12.5% x 0.90 = 11.25%	90.05%	10.33 ± 0.92%
C18:0	11.1% x 0.90 = 9.99%	90.05%	30.68 ± 6.56%

[1]Technical recovery from decarboxylation was 90.05%.

of C20:1 in E. comstockii reared on G. mellonella was
similar to that of G. mellonella which synthesizes the
acid, no incorporation of radioactivity occurred in that
acid in the parasite, suggesting direct deposition of
host fat into the parasite's lipids. The latter is, of
course, also the situation with the polyunsaturated fatty
acids, the levels of which are the same in parasite and
host, although they are not synthesized by either.

Further in vivo studies compared the incorporation of
radioactivity from labelled acetate when E. comstockii
was reared on two hosts (Barlow and Bracken, 1971). Despite marked differences in the relative levels of palmitic, palmitoleic and oleic acids in the parasite when
reared on alternate hosts, as expected from differences
in the fatty acid composition of the hosts, specific
activity differences indicated that direct dietary fat
deposition alone was not responsible. For example, the
level of palmitoleic acid in G. mellonella-reared and L.
sericata-reared E. comstockii differed by 10-fold, 2 and
20% respectively, but the specific activity was the same.
Specific activity may be considered as a ratio of the
amount of metabolite synthesized less the amount degraded
divided by the total present. Since this ratio for palmitoleic acid was the same regardlesss of a 10-fold difference in amount, the amount of synthesis less degradation must also have been changed 10-fold. Thus, the
level of palmitoleic acid appeared to be controlled by
metabolic processes. Similar results, however, did not
occur with all the fatty acids. For example, although
the level of palmitic acid was 10% higher (29% vs. 19%)
its specific activity was lower, suggesting some
deposition of dietary fat.

Additional studies with parasites were carried out

to determine the fatty acid composition of specific lipid fractions. In I. conquisitor the fatty acid compositions of both the phospholipid and neutral lipid fractions were considerably different when the parasite was reared on two hosts, G. mellonella and Ostrinia nubilalis (Thompson and Barlow, 1973). However, the phospholipid fraction, while adopting many of the characteristics of the host species, was not influenced to the same extent as the neutral lipid. In phospholipids, consistent ratios of the saturated and monounsaturated fatty acid classes were maintained, indicating that the parasite exerts some control over the physical nature of this lipid class although the composition is much more variable than in most animals. Studies with a second parasite, E. roborator, demonstrated that the fatty acid composition of the triglyceride and diglyceride were both qualitatively and quantitatively the same as the host species, but that the composition of the free fatty acid fraction was markedly different (Thompson and Adams, 1976). If regulation of esterified fatty acid levels was exerted during the synthesis and turnover of the fatty acids themselves, the free fatty acid fraction of the parasite should be similar to that of the host since the free fatty acid pool is considered the source of fatty acid substrate for complex lipid synthesis. Because this was not the case, it was suggested that acyltransferase specificity during glyceride synthesis may partially determine the fatty acid composition. The synthesis of all the major complex lipid classes including monoglyceride, diglyceride, and triglyceride from ^{14}C-2-acetate and the rapid incorporation of radiolabelled palmitic acid into esterified lipids of E. roborator has been demonstrated (Thompson and Johnson, 1978). The regulation of fatty acid levels through altered substrate specificity during glyceride synthesis is consistent with the finding of Barlow and Bracken (1971) described above. Following fatty acid synthesis, an increased acyltransferase specificity for one fatty acid in response to dietary alteration would result in an increased level of that fatty acid in the esterified lipids; the specific activity of that fatty acid would be unaltered. However, because the individual classes of esterified lipid all have different fatty acid compositions, the substrate specificity of the various acyltransferase systems must be affected by the proportion of fatty acids in each specific complex lipid. Recent studies have implicated additional metabolic factors in the "duplication" of host fatty acid composition.

RECENT ADVANCES

Further in vivo studies on lipid metabolism in parasitic Hymenoptera were facilitated by the development of

chemically-defined media for rearing the parasite Exeristes roborator under controlled nutritional conditions
(Thompson, 1975). E. roborator reared on fatty acid
free media rapidly synthesized fatty acids and, except
for the absence of polyunsaturates, the total body fats
had a fatty acid composition similar to that of most
animals in that palmitic and oleic acids predominated
while palmitoleic and stearic acids were present in
lesser amounts (Thompson and Barlow, 1976). This composition represents the true fatty acid synthetic capability of the parasite in the absence of regulating effects
of dietary fatty acids or of host metabolism. The distribution of radioactivity incorporated from ^{14}C-2-
acetate among the fatty acids was similar to that in
studies cited above with parasites reared on insect
hosts. E. roborator reared on fatty acid-free media
therefore retain a substantial capability for regulating
the fatty acid levels in the absence of dietary lipid.
However, supplementing the diets with purified lipids
from host insects had little influence on the composition
of the parasite. The differences in composition were
emphasized by very high MSD and MSDR values which were
orders of magnitude greater than those obtained with
that species reared on host material. Furthermore, the
composition of synthesized acids in the parasite was
similar to that observed when the parasite was reared on
fatty acid-free media. In contrast to these studies with
host lipid supplements, supplementing the diet with individual triglycerides altered the relative levels of the
corresponding fatty acids in E. roborator body fats
(Thompson and Johnson, 1978). In this case, however, the
increased levels were accompanied by decreases in the
relative specific activity suggesting direct dietary
deposition unaccompanied by increased de novo synthesis
of fat. For example, when diets were supplemented with
tripalmitolein the level of palmitoleic acid in the body
fats of E. roborator increased from 16 to 31%, approximately 2-fold, but the relative specific activity
decreased by half from 0.8 to 0.4. The difference in
the above results obtained with artificial media supplemented with host fat versus triglyceride supplements
remains unexplained, but neither result supports the suggestion that fatty acid synthesis or acyltransferase
specificity is altered in response to changes in dietary
fatty acid composition. It should be further noted, however, that because "duplication" was not observed in the
above studies with artificial media, both results are
inconsistent with those obtained in host reared insects
and described above.

The incorporation of radiolabelled precursors includ-
ing ^{14}C-2-acetate, ^{14}C-U-palmitate and ^{14}C(carboxyl)-tri-
palmitin into the various lipid classes of Exeristes

roborator reared on otherwise fatty acid-free artificial
media demonstrated rapid synthesis and interconversion
of complex lipid (Thompson and Johnson, 1978). However,
when the incorporation of ^{14}C-2-acetate was compared
with that observed in host-reared parasites it was found
that parasites reared on the artificial media synthesized
10 times more phospholipid in relation to neutral lipid
than those reared on G. mellonella and indeed synthesized
and stored very little triglyceride. Host-reared insects
contained twice the level of fat/unit wet weight as parasites reared on the artificial diet.

Recent in vitro studies demonstrated that despite
the high level of fat in host reared parasites, isolated
microsomes of E. roborator failed to incorporate glycerolphosphate into glycerides although they readily synthesized triglyceride by esterification using palmitic acid
(Table 5). In contrast, G. mellonella readily incorporated both glycerolphosphate and palmitic acid into phospholipid as well as triglyceride throughout its larval
development. Younger G. mellonella larvae incorporated
the greatest proportion of radioactivity into phospholipid while the older larvae incorporated it primarily
into triglyceride. In addition, the amount of glycerolphosphate incorporated into glyceride decreased as the
larvae grew older. These results indicate that the
younger G. mellonella larvae synthesize triglyceride and
phospholipid de novo, by esterification of glycerolphosphate, while the older larvae synthesize triglyceride by
esterification of partial glyceride. This conclusion
supports that of Peled and Tietz (1974), who suggested
that insects synthesize glyceride primarily from monoglyceride during fat mobilization and from glycerolphosphate
during growth. Garcia et al. (1980b) demonstrated that
larval and pharate adult C. capitata microsomal preparations synthesized triglyceride from monoglyceride at
equivalent rates, but larval preparations synthesized
glycerides by glycerolphosphate esterification and de
novo synthesis at much higher rates than preparations
from pharate adults resulting in a greater total amount
of glyceride synthesized in larvae. In contrast to G.
mellonella, host reared E. roborator synthesized glyceride principally via the partial glyceride pathways. In
this regard, it was of interest that preliminary in vivo
studies with E. roborator reared on fatty acid-free artificial media containing radioactive glycerolphosphate
demonstrated very little incorporation into triglyceride.

Most recently, similar studies with microsomal preparations from the parasite, I. conquisitor, which also
"duplicates" the fatty acid composition of its host,
demonstrated a lack of glycerolphosphate incorporation
into glycerides, while a number of other species including Brachymeria lasus, Aphaerites pallipes and Hyposoter

TABLE 5.
In vitro incorporation of ^{14}C-1-glycerolphosphate and ^{14}C-1-palmityl-CoA into microsomal lipids of the parasite _Exeristes roborator_ and its host, _Galleria mellonella_ (D. Jones and J. S. Barlow, unpublished).

Larvae		nmoles substrate incorporated/min/mg protein		Percent radioactivity incorporated from labelled palmitic acid			
Age (days)	Av. wt. (grams)	Glycerolphosphate	Palmitic acid	Phospholipid	Monoglyceride	Diglyceride	Triglyceride
E. roborator							
11	0.034	0	8.69	0	2	22	75
10	0.035	0	8.12	0	2	26	72
6	0.005	0	5.72	0	1	41	58
6	0.007	0	5.59	0	2	36	62
G. mellonella							
60	0.231	6.14	15.75	12	0	7	81
58	0.157	12.00	10.67	64	0	3	33
57	0.159	10.24	9.49	46	1	5	48
55	0.084	36.07	18.37	45	1	9	45
47	0.068	31.91	15.75	72	2	5	21

TABLE 6.
Relationship between the similarity of the fatty acid compositions
of parasitic Hymenoptera and their host species and the occurrence
of de novo glyceride synthesis (D. Jones, J. S. Barlow, and S. N.
Thompson, unpublished).

Parasite and Host	MSDR	Ratio of P-S-CoA incorporation into phospholipids/neutral lipid[1]
Family Braconidae Aphaereta pallipes Hylemya antiqua (Dip.)	0.71	0.54
Family Chalcidae Brachymeria lasus Trichoplusia ni (Lep.)	0.40	0.45
Family Ichneumonidae Exeristes roborator Pectinophora gossypiella (Lep.)	0.06	0
Itoplectis conquisitor Galleria mellonella (Lep.)	0.03	0
Hyposoter exiguae Trichoplusia ni (Lep.)	0.52	0.37

[1]This ratio reflects the relative occurrence of the de novo and monoglyceride pathways of glyceride synthesis because P-S-CoA is
incorporated into phospholipid by the de novo pathway only.

exiguae which do not have fatty acid compositions resembling their hosts, rapidly incorporated glycerolphosphate
(Table 6). Thus it appeared that a lack of de novo triglyceride synthesis may be characteristic of those parasites which "duplicate" the fatty acid composition of
their hosts, and that under nutritional conditions in
which dietary fats are present, the bulk of triglyceride
is synthesized by the esterification of partial glyceride
which presumably arises by lipid hydrolysis during digestion.

The above mechanism for triglyceride synthesis would
have a dramatic effect on the fatty acid composition of
the parasite's body fats. For example, if E. roborator
synthesized fatty acids in the proportions observed when
the parasite is reared on fatty acid-free artificial
media and when transferred to L. sericata, it synthesized
66% of its triglycerides from diglyceride and 33% from
monoglyceride of the new host (as indicated by the dis-

tribution of radioactivity into diglyceride and triglyceride from palmitic acid - Table 4) the resulting fatty acid composition when compared with the composition of that host would have an MSDR of 0.54, in contrast to an MSDR of 1.3 when the composition of the parasite reared on fatty acid-free media is compared directly to the host (since E. roborator reared on fatty acid-free media lacks polyunsaturates, the relative contribution of those fatty acids in L. sericata was added to the C16:1 level for calculation purposes). Although the MSDR value is substantially lower assuming the above mechanism is operative, it is nevertheless considerably higher than that obtained for this parasite reared on host material (Table 2). Furthermore, in evaluating the prevalence of the monoglyceride pathway as a contributing factor in "duplication", it is important to note that such a mechanism alone is not consistent with the results of Bracken and Barlow (1971) concerning the incorporation of radiolabelled precursors into lipids described above, nor with the analytical results of Thompson and Adams (1976) which demonstrated that the fatty acid compositions of the individual glyceride fractions are different although each of the compositions is the same as that of the corresponding glyceride of the host.

The interrelationship of de novo glyceride synthesis and the esterification of preformed partial glyceride to triglyceride are intimately involved in determining the fatty acid composition of animal fats but little is known about their regulation or the specific effects of nutritional status on their individual contributions to glyceride synthesis (Brindley, 1978). Dodds et al. (1976a) demonstrated in vitro in rat adipose tissue that partial glyceride inhibits glycerolphosphate esterification and similar findings were reported by Garcia et al. (1980b) in microsomal preparations of C. capitata larvae suggesting that nutritional factors play a role in regulating the activity of these systems in the intact animal. Dodds et al. (1976b) later determined the effects of high carbohydrate and fat diets on the activities of glycerolphosphate acyltransferase and monoacylglycerolphosphate acyltransferase in rats, but the results did not appear to correlate well with the in vitro effects.

In contrast to the above in vitro studies with microsomal preparations additional in vivo studies demonstrated that E. roborator can synthesize glyceride de novo and that nutrition does play an important role in the lipogenic processes of the parasite. When this species was reared on fatty acid-free media the level of total body fat was strongly influenced by dietary carbohydrate (Thompson, 1979a). Larvae reared on diets lacking carbohydrate contained very little fat although they incorporated radioactive glycerolphosphate into phospholipid.

However, inclusion of glucose increased the rate of lipogenesis sharply and the resulting increase in lipid content was largely triglyceride. Since the media were fat-
free, this glyceride must be synthesized de novo as indicated by the increased incorporation of radioactive
glycerolphosphate into triglyceride when compared to that
observed in animals reared on carbohydrate-free media.
Furthermore, the fatty acid component of the glyceride
must be synthsized de novo as well, and Thompson (1979b)
demonstrated in vitro that fatty acid synthetase activity
was correlated to dietary carbohydrate level. Short et
al. (1977) previously demonstrated in rabbits a close
parallel between the activities of glycerolphosphate
acyltransferase and fatty acid synthetase. Recent in
vitro studies with microsomal preparations of E. roborator reared on fatty acid-free artificial media containing
high levels of glucose have demonstrated low levels of
glycerolphosphate incorporation into glyceride, but
unfortunately this was not quantitatively correlated to
the in vivo incorporation (Jones, Barlow and Thompson,
unpublished). Dietary carbohydrate also had a marked
effect on fatty acid composition in E. roborator (Thompson, 1979b). The fatty acids of parasites reared on carbohydrate-free media were composed primarily of saturated
acids, which may reflect a low level of de novo fatty
acid synthesis and result in a lack of saturated substrate for desaturation. The dietary effects on saturated fatty acid synthesis were demonstrated in vitro and
the activity of isolated fatty acid synthetase was found
to be proportional to the dietary glucose level. The
proportion of unsaturated fatty acids in both the phospholipid and triglyceride fractions increased with increasing dietary carbohydrate levels. However, these
alterations represent the effects of dietary glucose on
the overall activity of lipogenic enzymes. To accurately
determine how dietary carbohydrate influences fatty acid
composition, the rates of change in the absolute amount
of saturated and unsaturated fatty acids in the phospholipids and triglycerides synthesized were calculated.
In most cases, the rates increased in response to dietary
glucose levels. Of interest, however, were the low or
negative rates for the saturated fatty acids in the phospholipid fraction indicating that the absolute amount of
saturated fatty acids in this fraction decreased in response to dietary carbohydrate. Since this was not the
case in the triglycerides, these lower rates are likely
the result of altered esterification patterns during glyceride synthesis. The increase in saturated fatty acid
rate for the triglyceride fraction which accompanied the
decreased rate in phospholipid was presumably due to a
shift of fatty acids into triglyceride.

Regarding glyceride metabolism in other groups of

metazoan parasites, several studies have demonstrated the rapid incorporation of glycerol, fatty acids and phosphorous into the complex lipids of parasitic helminths (Meyer and Meyer, 1972). Indeed, the de novo pathway has been identified in Hymenolepis diminuta, but the enzymes have not been characterized (Buteau and Fairbairn, 1969). Despite the fact that selective acylation has been suggested as the mechanism of control over fatty acid composition in the complex lipids little is known about the specific nature of such selectivity. Buteau and Fairbairn (1969), however, have demonstrated an asymmetric distribution of fatty acids in the triglycerides of H. diminuta with saturated fatty acids primarily at position 1, monounsaturated acids at positions 1 and 3 and polyunsaturates at position 2. Because of this non-random distribution the monoglyceride pathway was suggested to be absent. Although direct evidence for this lack in capacity was not obtained, H. diminuta was previously shown to rapidly hydrolyze absorbed 2-monoglyceride, and thus, the monoglyceride pathway may be of little quantitative importance in any case (Bailey and Fairbairn, 1968). Lysophosphatidyl choline, another major dietary lipid for H. diminuta, was also freely absorbed but appeared to be hydrolyzed prior to glyceride synthesis.

SUMMARY

To date, several aspects of lipid metabolism in parasitic Hymenopterta have been examined. Specifically, studies have been aimed at characterizing the lipid composition including the fatty acid composition of the complex lipids, determining the nature of fatty acid and complex lipid synthesis as well as the effects of nutritional status on these compositions and metabolic processes. The investigations have demonstrated the unique ability of many parasite species to regulate their fatty acid composition in a fashion which results in a near identical composition to their host species. A number of specific metabolic factors contribute in a complex fashion to the ultimate determination of the parasite's fatty acid levels. These appear to include fatty acid turnover, the substrate specific nature of the acyltransferase enzymes involved in complex lipid formation as well as the balance between the de novo and monoglyceride pathways of complex lipid synthesis. Furthermore, the above factors are strongly influenced by dietary carbohydrate and lipid, but no single factor has been observed as the determining one under any specific instance thus far examined. Further studies on the effects of nutritional status on the above metabolic processes will hopefully result in a clearer picture of the specific nature of regulation involved in the "duplication" phenomenon.

CONCLUSION

The metabolic determination of fatty acid composition
in the complex lipids of animal fats involves several
enzyme complexes and multitudes of individual enzymes.
Fatty acid composition is affected dramatically by environmental conditions, nutritional state, and undoubtedly,
hormones, a factor not considered in the present discussion. In most animals the ultimate effect of these factors on the balance of the metabolic pathways which determine the fatty acid composition, appear similar and
results in the maintenance of a characteristic fatty acid
composition. The situation in some of the Hymenoptera,
however, is unique. Although the basic mechanisms of
fatty acid and glyceride synthesis are intact, nutritional effects appear more pronounced than in most animals,
which may enable the organism to take maximum benefit
from its limited food sources. For example, an extreme
depression of de novo glyceride synthesis under a high
fat nutritional regime would allow the parasite to use
its host's fats after partial digestive hydrolysis and
its own fatty acids for rapid triglyceride synthesis,
thereby minimizing the energy cost of fat synthesis.
The absence of de novo fatty acid biosynthesis in parasitic helminths might have a similar function although
stored fat does not appear to be important in the energy
metabolism of those species examined and anaerobic carbohydrate utilization rather than fat mobilization is typical. Since adaptation is the result of natural selection, and is thus goal-directed, any suggested adaptive
value of fat synthesis in parasitic helminths must emphasize the fact that triglycerides have no known function
in many of these animals. Similarly, in the Hymenoptera,
the biological and/or physiological significance of altered fatty acid synthesis and/or acyltransferase specificity and the resultant "duplicate" fatty acid composition remain perplexing. The mechanism involved may be a
result of accidental selection. The latter, while gene-
tically associated with adaptation and perhaps an important factor in evolution, is not goal-directed and thus
it is not always possible to determine the significance
of traits which arise in this manner. On the other hand,
since it may be more physiologically accurate to consider
adaptation as the genetic capacity for responding to environmental pressure rather than the response itself,
certain adaptations are not always readily observable,
and such may be the case here. Further study is necessary to determine the mechanisms which regulate the fatty
acid composition in parasitic Hymenoptera and to assess
their genetic basis and possible biological significance.

REFERENCES

Ailhaud, G., D. Samuel, M. Lazdunski, and P. Desnuelle. (1964) Quelques observations sur le mode d'action de la monoglyceride transacylase et de la diglyceride transacylase de la muqueuse intestinale. Biochim. Biophys. Acta 84: 643-664.

Askew, R.R. (1971) Parasitic insects, ix-xvii. Elsevier, New York, 316 pp.

Bailey, H. H. and D. Fairbairn. (1968) Lipid metabolism in helminth parasites. V. Absorption of fatty acids and monoglycerides from micellar solution by Hymenolepis diminuta (Cestoda). Comp. Biochem. Physiol. 26: 819-836.

Barlow, J.S. (1965) Effects of diet on the composition of body fat in Agria affinis. Can. J. Zool. 43: 337-340.

Barlow, J.S. (1966a) Effects of diet on the composition of body fat in Musca domestica. Can. J. Zool. 44: 775-779.

Barlow, J.S. (1966b) Effects of diet on the composition of body fat in Lucilia sericata. Nature 212: 1478-1479.

Barlow, J.S. and G. K. Bracken. (1971) Incorporation of Na-1-^{14}C-acetate into the fatty acids of two insect parasites (Hymenoptera) reared on different hosts. Can. J. Zool. 49: 1297-1300.

Barlow, J.S. (1972) Some host-parasite relationships in fatty acid metabolism. In Insect and mite nutrition, pp. 437-452 (J.G. Rodriguez, ed., North-Holland Publishing Co., New York).

Barlow, J.S. and D. Jones. (1980) Inhibition of the use of endogenous substrates by Galleria mellonella transacylases. Can. J. Zool. 58: 598-608.

Barlow, J.S., S.N. Thompson, and J. Martini. (1980) Incorporation of L-α-glycerolphosphate and acyl-S-CoA into lipids by microsomes of Galleria mellonella. Can. J. Zool. 58: 50-56.

Barnett, J.W. and R.S. Berger. (1970) Dietary fats and fatty acid composition of bollworms as factors influencing insecticide susceptibility. J. Econ. Ent. 63: 1430-1433.

Bracken, G.K. and J.S. Barlow. 1967. Fatty acid composition of Exeristes comstockii (Cress.) reared on different hosts. Can. J. Zool. 45: 57-61.

Bridges, R.G. and S.G. Watts. (1975) Changes in fatty acid composition of phospholipids and triglycerides of Musca domestica resulting from choline deficiency. J. Insect Physiol. 21: 861-871.

Brindley, D.N. (1978) Regulation of fatty acid esterification in tissues. In Regulation of fatty acid and glycerolipid metabolism, 31-40 (R. Dils and J. Knudsen, ed., Pergamon Press, New York).

Brindley, D. N. and G. Hübscher. (1966) The effect of chain length on the activation and subsequent incorporation of fatty acids into glycerides by the small intestinal mucosa. Biochim. Biophys. Acta 125: 92-105.

Buteau, G.H. and D. Fairbairn. (1969) Lipid metabolism in helminth parasites. VIII. Triglyceride synthesis in Hymenolepis diminuta (Cestoda). Exp. Parasitol. 25: 205-275.

Caster, W.O., H. Mohrhauer, and R.T. Holman. (1966) Effects of twelve common fatty acids in the diet upon the composition of liver lipid in the rat. J. Nutrition 89: 217-225.

Clark, B. and G. Hübscher. (1963) Monoglyceride transacylase of rat intestinal mucosa. Biochim. Biophys. Acta 70: 43-52.

Dodds, P.F., M.I. Gurr, and D.N. Brindley. (1976a) The glycerol phosphate, dihydroxyacetone phosphate and monoacylglycerol pathways of glycerolipid synthesis in rat adipose tissue homogenates. Biochem. J. 160: 693-700.

Dodds, P.F., D.N. Brindley, and M.I. Gurr. (1976b) The effects of diet on the esterification of glycerol phosphate, dihydroxyacetone phosphate and 2-hexadecylglycerol by homogenates of rat adipose tissue. Biochem. J. 160: 701-706.

Downer, R.G.H. and J.R. Mathews. (1976) Patterns of lipid distribution and utilization in insects. Am. Zool. 16: 733-746.

DeKruyff, B., P.W.M. Van Dyck, R.W. Goldbach, R.A. Demel, and L.L.M. Van Deenan. (1973) Influence of fatty acid and sterol composition on the lipid phase transition and activity of membrane bound enzymes in Acholeplasma laidlawaii. Biochim. Biophys. Acta 330: 269-282.

Dwivedy, A.K. (1977) Effects of cholesterol deficiency on the composition of phospholipids and fatty acids of the housefly, Musca domestica. J. Ins. Physiol. 23: 549-557.

Fairbairn, D. (1970) Biochemical adaptation and loss of genetic capacity in helminth parasites. Biol. Rev. 45: 27-72.

Garcia, R. and A.M. Municio. (1980) Incorporation of fatty acids and the positional distribution analysis of fatty acids in acylglycerols of the insect Ceratitis capitata. Comp. Biochem. Physiol. 66B: 195-203.

Garcia, R., A. Megias, and A.M. Municio. (1980a) Biosynthesis of neutral lipids by mitochondria and microsomes during development of insects. Comp. Biochem. Physiol. 65B: 13-24.

Garcia, R., A.M. Municio, and D. Viloria. (1980b) Glycerol phosphate and monoacylglycerol pathways of triacylglycerol biosynthesis in microsomes of Ceratitis capitata. Comp. Biochem. Physiol. 66B: 435-437.

Gonzales-Buitrago, J. M., A. Megias, A. M. Municio, and M.A. Perez-Albarsanz. (1979) Fatty acid elongation and unsaturation by mitochondria and microsomes during development of insects. Comp. Biochem. Physiol. 64B: 1-10.

Groot, P.H.E., H.R. Scholte, and W.C. Hülsmann. (1976) Fatty acid activation: specificity, localization and function. Adv. Lipid Res. 14: 75-126.

Hoffman, A.G.D. and R.G.H. Downer. (1979) Synthesis of diacylglycerides by monoacylglycerol acyltransferase from crop, midgut and fat body tissues of the American cockroach, Periplaneta americana L. Ins. Biochem. 9: 129-134.

House, H. L., D. F. Riordan, and J. D. Barlow. (1958) Effects of thermal conditioning and of degree of saturation of dietary lipids on resistance of an insect to a high temperature. Can. J. Zool. 36: 629-632.

Howling, D., L. J. Morris, M. I. Gurr, and A. T. James. (1972) The specificity of fatty acid desaturation and hydroxylases. Biochim. Biophys. Acta 260: 10-19.

Hübscher, G. (1969) Glyceride metabolism. In Lipid metabolism, 280-370 (S.J. Wakil, ed., Academic Press, New York).

Johnson, J.M., G.A. Rao, and P.A. Lowe. (1967) The separation of the glycerophosphate and monoglyceride pathways in the intestinal biosynthesis of triglycerides. Biochim. Biophys. Acta 137: 578-580.

Keith, A.D. (1966) Analysis of lipids in Drosophila melanogaster. Comp. Biochem. Physiol. 17: 1127-1139.

Knipprath, W.G. and J.R. Mead. (1968) The effect of environmental temperature on the fatty acid composition and on the in vivo incorporation of 1-^{14}-C-acetate in goldfish (Carassius auratus L.). Lipids 3: 121-138.

Kuhn, N.J. (1967) Esterification of glycerol-3-phosphate in lactating guinea-pig mammary gland. Biochem. J. 105: 213-223.

Lands, W.E.M. and P. Hart. (1965) Metabolism of glycerolipids. VI. Specificities of acyl coenzyme A: phospholipid acyltransferases. J. Biol. Chem. 240: 1905-1911.

McCaman, R.F., M. Smith, and K. Cook. (1965) Intermediary metabolism of phospholipids in brain tissue. II. Phosphatidic acid phosphatase. J. Biol. Chem. 240: 3513-3517.

Megias, A., A. M. Municio, and M. A. Perez-Albarsanz. (1977) Biochemistry of development in insects. Triacylglycerol and phosphoglyceride biosynthesis by subcellular fractions. Eur. J. Biochem. 72: 9-16.

Meyer, F. and H. Meyer. (1972) Loss of fatty acid bio-
    synthesis in flatworms. In Comparative biochemistry
    of parasites, 383-393 (H. Van den Bossche, ed.,
    Academic Press, New York).
Moore, R.F. (1980) The effect of varied amounts of
    starch, sucrose, and lipids on the fatty acids of the
    boll weevil. Ent. Exp. Appl. 27: 246-254.
Municio, A.M., J.M. Odriozola, A. Pineiro, and A. Ribera.
    (1971) In vitro fatty acid and lipid biosynthesis
    during development of insects. Biochim. Biophys.
    Acta 248: 212-225.
Municio, A.M., J.M. Odriozola, A. Pineiro, and A. Ribera.
    (1972) In vitro elongation and desaturation of fatty
    acids during development of insects. Biochim. Bio-
    phys. Acta 280: 248-257.
Municio, A.M., J.M. Odriozola, M.A. Perez-Albarsanz, and
    J.A. Ramos. (1974) In vitro and in vivo fate of
    saturated and unsaturated fatty acids during develop-
    ment of insects. Biochim. Biophys. Acta 360:289-297.
Municio, A.M., J.M. Odriozola, and M.A. Perez-Albarsanz.
    (1975a) Biochemistry of development in insects.
    Incorporation of fatty acids into different lipid
    classes. Eur. J. Biochem. 60: 123-128.
Municio, A.M., R. Garcia, and M.A. Perez-Albarsanz.
    (1975b) Biochemistry of development in insects.
    Stereospecific incorporation of fatty acids into
    triacylglycerols. Eur. J. Biochem. 60: 117-121.
Municio, A.M., M.A. Lizarbe, E. Relano, and J.A. Ramos.
    (1977) Fatty acid synthetase complex from the insect
    Ceratitis capitata. Biochim. Biophys. Acta 487: 175-
    188.
Normann, P.T. and T. Flatmark. (1980) Acyl-CoA synthe-
    tase activity of brown adipose tissue mitochondria.
    Substrate specificity and its relation to the endo-
    genous pool of long-chain fatty acids. Biochim.
    Biophys. Acta 619: 1-10.
Peled, Y. and A. Tietz. (1974) Acylation of monoglycer-
    ides by locust fat body microsomes. FEBS lett. 41:
    65-68.
Possmayer, E., G. L. Scherphof, M. A. R. Dubbelman, and
    L.L.M. Van Deenen. (1969) Positional specificity
    of saturated and unsaturated fatty acids in phospha-
    tidic acid from rat liver. Biochim. Biophys. Acta
    176: 95-110.
Pynadath, T.I. (1969) The mechanism of the chain elon-
    gation of fatty acids by soluble extracts of beef
    liver mitochondria. Fed. Proc. 28: 537.
Rao, G.A. and J.M. Johnson. (1966) Purification and
    properties of triglyceride synthetase from the intes-
    tinal mucosa. Biochim. Biophys. Acta 125: 465-473.
Schaefer, C.H. (1968) The relationship of the fatty
    acid composition of Heliothis zea larvae to that of

its diet. J. Insect Physiol. 14: 171-178.

Short, V.J., D.N. Brindley, and R. Dils. (1977) Coordinate changes in enzymes of fatty acid synthesis, activation and esterification in rabbit mammary glands during pregnancy and lactation. Biochem. J. 162: 445-450.

Shortland, F.B. (1952) Evolution of animal fats. Nature 170: 924-925.

Strong, F.E. (1963) Studies on lipids in homopterous insects. Hilgardia 34: 43-61.

Takaya, T. and K. Miura. (1968) Biosynthesis of fatty acids in the larvae of the blowfly, Aldrichina grahami, reared aseptically. Arch. Inter. Physiol. Biochem. 76: 603-614.

Tame, M.J. and R. Dils. (1969) Factors controlling the chain length of fatty acids synthesized by the intestinal mucosa of guinea-pig. Experimentia 25: 21-22.

Taylor, F.R. and L.W. Parks. (1979) Triacylglycerol metabolism in Saccharomyces cerevisiae: relation to phospholipid synthesis. Biochim. Biophys. Acta 575: 204-214.

Thompson, S.N. (1973) A review and comparataive characterization of the fatty acid composition of seven insect orders. Comp. Biochem. Physiol. 45: 467-482.

Thompson, S.N. (1975) Defined meridic and holidic diets and aseptic feeding procedures for artificially rearing the ectoparasitoid, Exeristes roborator (Fabricius). Ann. Entomol. Soc. Amer. 68: 220-226.

Thompson, S.N. (1979a) Effects of dietary carbohydrate on larval development and lipogenesis in the parasite, Exeristes roborator (Fabricius) (Hymenoptera: Ichneumonidae). J. Parasitol. 65: 849-854.

Thompson, S.N. (1979b) Effect of dietary glucose on in vivo fatty acid metabolism and in vitro synthetase activity in the insect parasite, Exeristes roborator (Fabricius). Ins. Biochem. 9: 645-651.

Thompson, S.N. (1981) Nutrition of parasitic Hymenoptera. Proceedings of the IX Inter. Congr. of Plant Protec. 1: 93-96.

Thompson, S.N. and J.D. Adams. (1976) Characterization of selected lipids of the parasite Exeristes roborator (Fabricius). Comp. Biochem. Physiol. 55B: 591-593.

Thompson, S.N. and J.S. Barlow. (1970) The change in fatty acid pattern of Itoplectis conquisitor (Say) reared on different hosts. J. Parasitol. 56: 845-846.

Thompson, S.N. and J.S. Barlow. (1971) Aspects of fatty acid metabolism in Galleria mellonella. Comp. Biochem. Physiol. 38B: 333-346.

Thompson, S.N. and J.S. Barlow. (1972a) The consistency of the fatty acid pattern of Galleria mellonella

106

reared on fatty acid supplemented diets. Can. J. Zool. 50: 1033-1034.

Thompson, S.N. and J.S. Barlow. (1972b) Synthesis of fatty acids by the parasite Exeristes comstockii (Hymenoptera) and two hosts, Galleria mellonella (Lepidoptera) and Lucilia sericata (Diptera). Can. J. Zool. 50: 1105-1110.

Thompson, S.N. and J.S. Barlow. (1973) The inconsistent phospholipid fatty acid composition of an insect parasitoid, Itoplectis conquisitor. Comp. Biochem. Physiol. 44B: 59-64.

Thompson, S.N. and J.S. Barlow. (1974) The fatty acid composition of parasitic Hymenoptera and its possible biological significance. Ann. Entomol. Soc. Amer. 67: 627-632.

Thompson, S.N. and J.S. Barlow. (1976) Regulation of lipid metabolism in the insect parasite, Exeristes roborator. J. Parasitol. 62: 303-306.

Thompson, S.N., J.S. Barlow, and V.M. Douglas. (1975) Preliminary purification and properties of a fatty acid synthetase complex isolated from the blowfly, Lucilia sericata. Ins. Biochem. 5: 571-583.

Thompson, S.N. and J. Johnson. (1978) Further studies on lipid metabolism in the insect parasite Exeristes roborator (Fabricius). J. Parasitol. 64: 731-740.

Tietz, A. (1969) Biosynthesis of diglycerides and triglycerides in cell free preparations of the fat body of the locust. Israel J. Med. Sci. 5: 1007-1017.

Teitz, A. and N. Stein. (1969) Stearate desaturation by microsomes of the locust fat body. FEBS lett. 2: 286.

Tietz, A., H. Weintraub, and Y. Peled. (1975) Utilization of 2 acyl-sn-glycerol by locust fat body microsomes: specificity of the acyltransferase system. Biochim. Biophys. Acta 388: 11-21.

Vanderzant, E.S. (1968) Dietary requirements of the bollworm, Heliothis zea for lipids, choline and inositol and the effects of fats and fatty acids on the composition of body fat. Ann. Entomol. Soc. Amer. 61: 120-125.

Vinson, S.B. 1975. Biochemical coevolution between parasitoids and their hosts. In Evolutionary strategies of parasitic insects and mites, 14-48 (P. Price, ed., Plenum Press, New York, 225 pp.).

Vinson, S.B. (1976) Host selection by insect parasitoids. Ann. Rev. Ent. 21: 109-133.

Vinson, S.B. and G.F. Iwantsch. (1980) Host suitability for insect parasites. Ann. Rev. Ent. 25: 297-419.

Wilson, G., S.P. Rose, and C.F. Fox. (1970) The effect of membrane lipid unsaturation on glycoside transport. Biochem. Biophys. Res. Commun. 30: 617-623.

7
Essential Fatty Acids: Insects and Vertebrates Compared

R. H. Dadd*

Evidence that vertebrates might require dietary poly-unsaturated fatty acids first came from work with rats in the late 1920s. Though the nature of essential lipid factors in highly unsaturated oils was sometimes in dispute, investigations with several vertebrates over the next two decades led to a concensus that polyunsaturates were the active substances (Sinclair, 1964; Holman, 1967). Since then, extensive research has greatly clarified metabolic relationships between the dietary and physiologically essential fatty acids of vertebrates and has indicated their several physiological functions. A dietary requirement for polyunsaturated fatty acids in an insect was first revealed by studies on flour moth nutrition (Fraenkel and Blewett 1946). Similar needs have since been demonstrated for many insects of diverse orders and dietetic habit, but even now there is virtually no metabolic information bearing on the crucial biochemical or physiological functions of the essential fatty acids in insects.

Recently it was found that the dietary polyunsaturated fatty acids needed by mosquitoes differed from those able to satisfy the requirement of all insects previously studied. On the other hand, fatty acids that satisfy the mosquito requirement include those that are of central physiological importance for vertebrates. This situation raises many questions about insect fatty acid requirements in general and hints at parallels between vertebrate and insect essential fatty acid requirements in terms of ultimate physiological function. Since the essential fatty acid requirements and metabolism of vertebrates are now well understood they provide a convenient framework against which to consider the relative-

* Department of Entomology, University of California, Berkeley, CA 94720, USA

ly fragmentary information about insects. Current orthodoxy with respect to vertebrates is therefore first summarized so as to lend some order to interpretation of the
primarily nutritional information available for insects.

VERTEBRATE ESSENTIAL FATTY ACIDS

This synopsis draws mainly on the reviews of Mead
(1970), Guarnieri and Johnson (1970), Alfin-Slater and
Aftergood (1971), Holman (1977), and Tinoco et al.
(1979). In warm-blooded vertebrates the gross pathologies of essential fatty acid deficiency include reduced growth, epidermal lesions, increased water loss
through the skin resulting in increased drinking, and
infertility in both males and females. Various subcellular lesions have been observed to precede or accompany
the gross pathologies; these lesions mostly affect lipid

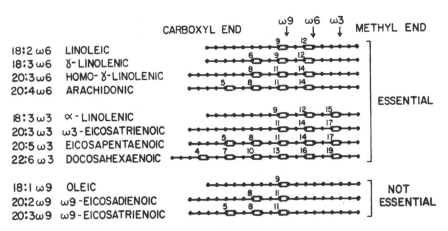

FIGURE 1. Structural features of the carbon chain of
essential fatty acids.

membranes of cells and their organelles, causing changes in membrane permeability under certain conditions of stress, and altered kinetics of membrane-bound enzymes. Biochemically, deficiency is characterized by a reduction in the levels of the tetraene, arachidonic acid ($\Delta 5,8,11,14$-C20:4 or 20:4ω6) in tissue lipids, most marked in the phospholipids, which predominate in organs such as nerves and glands having a high component of lipid membranes. The decline in arachidonic acid tends to be accompanied by an increase in polyunsaturates derived from oleic acid ($\Delta 9$-C18:1 or 18:1ω9), especially the eicosatrienoic acid, $\Delta 5,8,11$-C20:3 (or 20:3ω9), leading to an increase in the triene/tetraene ratio.

In warm-blooded vertebrates, all deficiency symptoms can be remitted by dietary fatty acids of the ω6 (or n6) family, whose features are indicated by the diagrams of Figure 1 and whose root member is the common plant diene, linoleic acid ($\Delta 9,12$-C18:2 or 18:2ω6). Linoleic acid can be metabolized by carbon-chain elongation and further desaturation to arachidonic acid, as shown in Figure 2, and all intermediates as well as linoleic and arachi-

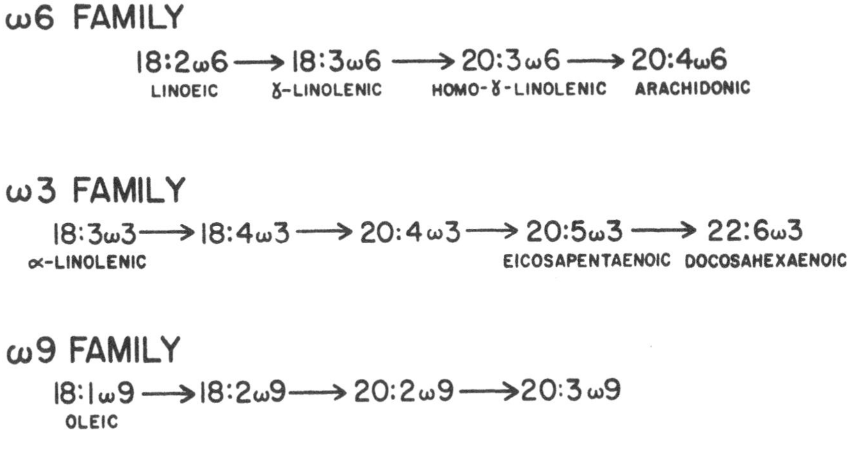

ω6 FAMILY

$$18:2\omega6 \longrightarrow 18:3\omega6 \longrightarrow 20:3\omega6 \longrightarrow 20:4\omega6$$

LINOEIC γ-LINOLENIC   HOMO-γ-LINOLENIC   ARACHIDONIC

ω3 FAMILY

$$18:3\omega3 \longrightarrow 18:4\omega3 \longrightarrow 20:4\omega3 \longrightarrow 20:5\omega3 \longrightarrow 22:6\omega3$$

α-LINOLENIC   EICOSAPENTAENOIC   DOCOSAHEXAENOIC

ω9 FAMILY

$$18:1\omega9 \longrightarrow 18:2\omega9 \longrightarrow 20:2\omega9 \longrightarrow 20:3\omega9$$

OLEIC

FIGURE 2. Major metabolic pathways for unsaturated fatty acids of the ω6, ω3 and ω9 families in mammals.

donic acids are fully effective in remitting essential fatty acid deficiency. Many, though not all, symptoms and indices of deficiency can also be ameliorated by the common plant triene, α-linolenic acid (Δ9,12,15-C18:3 or 18:3ω3), the root member of the ω3 (or n3) family of fatty acids. As with linoleic acid, linolenic acid may be elongated and further desaturated through a series of intermediates up to docosahexaenoic acid (Δ5,7,10,13,16, 19-C22:6 or 22:6ω3), as diagrammed in Figure 2, and these members also are effective in alleviating many deficiency symptoms.

The enzymes that metabolize linoleic and linolenic acids to higher members of their respective families act preferentially on linoleic acid, linolenic acid and the lower members of the ω6 and ω3 families, so if these fatty acids are available from the diet, their higher members predominate among tissue long-chain polyunsaturates. However, in their absence, oleic acid and other members of the ω9 family assume the substrate role, leading to production of the aforementioned unusual ω9 eicosatrienoic acid (see Figure 2) and to the high triene/tetraene ratio diagnostic of essential fatty acid deficiency.

Current orthodoxy holds that the dietary requirement for essential fatty acids reflects two main physiological roles: a structural need for certain polyunsaturates in the molecules of specific membrane phospholipid species, thought to optimize certain membrane-associated functions; and a need for precursors of prostaglandins, thromboxanes and leucotrienes, comparatively recently delineated hormone-like entities, some or other of which are ubiquitously present in physiologically active vertebrate tissues and are involved in the regulation of a wide array of localized or cellular metabolic processes.

With respect to the membrane structural role, the relative abundance of polyunsaturated fatty acids found in phospholipids in relation to the availability of ω6, ω3 or ω9 substrates for desaturation suggests a hierarchy of structural specificities in lipid membrane-dependent functions: for some functions, only arachidonic acid may suffice; for others, perhaps various fatty acids of a particular chain length and/or degree of unsaturation, including both ω6 and ω3 members, may be interchangeably adequate; and lowest in the hierarchy may be non-fastidious functions requiring merely an optimization of overall unsaturation for the total phospholipid fatty acid complement, achievable by diverse combinations of polyunsaturates.

With respect to the prostaglandin precursor function, the required fatty acid specificity is narrower, since only 20-carbon polyunsaturates are suitable as immediate precursors: homo-γ-linolenic acid (20:3ω6) and arachidonic acid ($20:4\omega6$) for the prostaglandin 1 and 2 series, respectively; and eicosapentaenoic acid ($20:5\omega3$) for the prostaglandin 3 series (Lands et al. 1977). Most physiologically active prostaglandins of mammals are in the 1 and especially the 2 series which derive specifically from $\omega6$ fatty acids, predominately arachidonic acid, which latter is also the precursor of the very recently discovered leucotrienes (Lewis and Austen 1981). This has led to the suggestion that all crucial physiological functions of essential fatty acid may reflect an ultimate need for prostanoids of various sorts. Were this the case, it would follow that suitability as a substrate for prostaglandinogenesis defines the essentiality of the fully essential fatty acids. One can then visualize that partial essentiality might reflect adequacy for a structural membrane role only, with degrees of adequacy reflecting the extent of fatty acid structural suitability for possibly several lipid membrane functions.

In contrast to the primacy of arachidonic acid in the essential fatty acid physiology of mammals and birds, α-linolenic acid or higher members of the $\omega3$-family are the primary essential fatty acids for many fish. Indeed, fish triglycerides as well as phospholipids, characteristically contain very high proportions of $\omega3$ polyunsaturates such as eicosapentaenoic and docosahexaenoic acids, perhaps an adaptation to maintain lipid fluidity at the relatively low body temperature of aquatic poikilotherms.

For both fish and warm-blooded vertebrates it is uncertain whether or not the secondary essential fatty acid family, $\omega3$ in the case of mammals and $\omega6$ for fish, may also be independently and additionally essential. Some evidence points to a dietary need for linolenic as well as linoleic acid to maintain a normal brain phospholipid fatty acid composition in mammals (Galli et al. 1976). And if fish depend upon prostaglandins of the 1 and 2 series, this would imply the essentiality of linoleic or other $\omega6$ fatty acids.

It is generally assumed that the essentiality of certain fatty acids reflects an inability to biosynthesize them de novo or by conversion from those saturated and monoenoic acids that can be synthesized by an animal. The overwhelming evidence from vertebrate metabolic studies indicates that fatty acids with more than one double bond cannot be biosynthesized, nor do conversions between members of the $\omega3$, $\omega6$, and $\omega9$ families occur. This, and a preponderance of similar findings from metabolic studies with invertebrates, lends authority to the widely held generalization that animals cannot biosynthesize polyunsaturated fatty acids. In a later section the validity of this generalization with respect to insects is considered.

ESSENTIAL FATTY ACIDS FOR INSECTS

The earliest work on insect essential fatty acids, that of Fraenkel and Blewett (1946) with flour moths of the genus Ephestia, still serves to illustrate the main features of deficiency as they were understood up to a few years ago. With casein-based meridic diets lacking fatty acid, larval growth of Anagasta (Ephestia) kuehniella was retarded, and though pupation usually occurred, adult moths failed to emerge properly. Certain highly unsaturated vegetable oils alleviated these symptoms, but neither saturated fats, vegetable or animal, nor highly unsaturated fish oil, supported normal adult emergence, though fish oil, and to some extent olive oil, improved the larval growth rate. All beneficial effects of unsaturated vegetable oil were obtained by substituting either linoleic or linolenic acids for oil, linolenic acid being perhaps the more potent. Docosahexaenoic acid (22:6ω3), a long-chain highly polyunsaturated fatty acid characteristic of fish oil, failed to alleviate pupal/adult failure, though, like fish oil, it improved larval growth. Oleic acid and arachidonic acid likewise failed to support normal adult emergences though both speeded the larval growth rate to some extent.

Similar findings were obtained with approximately 60% out of all species for which nutritional effects of polyunsaturated fatty acids were sought (Dadd 1973, 1977, 1981), though the deficiency was not always expressed so dramatically at the pupal/adult moult. Failure at adult emergence characterizes the deficiency for most Lepidoptera, sometimes with, and sometimes without preceding larval growth retardation. Similar adult malformations or failure occur in fatty acid-deficient acridid grasshoppers (Dadd 1961; Nayar 1964) and certain parasitoid Hymenoptera (Yazgan 1972; Thompson 1981), sometimes with preceding larval growth retardation, and are also characteristic of fatty acid deficiency in mosquitoes, as discussed in more detail in a later section. However, with some insects, such as certain Coleoptera (Vanderzant and Richardson 1964; Earle et al., 1967; Wardojo 1969), and the cockroach Blattella germanica (Gordon 1959), fatty acid deficiency manifests only as delayed larval growth or reproductive decline, sometimes requiring more than one generation of deprivation to become apparent.

In work dating mainly from 4 decades ago several stored products beetles and the dozen or so Diptera then studied nutritionally using synthetic diets appeared not to require dietary fatty acid for growth over one larval cycle, and indeed 3 of the 20 or more Lepidoptera studied more recently have shown no requirement (e.g., Chippendale 1971). In spite of this, it has been argued that a need for essential unsaturated fatty acid is probably universal among insects (Dadd 1973, 1979, 1981; Dadd and Kleinjan 1979), a speculation based on the following considerations. First, fatty acid deficiency characteristically becomes evident only late in development, as summarized above. In some Lepidoptera the pupal/adult failure is entirely averted by feeding an essential fatty acid during the final larval stadium only, suggesting the physiological need for fatty acid becomes critical only with the advent of metamorphosis and adult maturation (Rock et al., 1965; Grau and Terriere 1971). Thus a requirement might easily remain undetected in the usual single generation growth experiment. On this view, negative findings with respect to a fatty acid requirement may be indeterminate. In this connection it is of interest that another flour moth, Plodia interpunctella, which apparently did not require dietary fatty acid in the original study of Fraenkel and Blewett (1946), has since been shown to require dietary fatty acids for proper emergence and adult fertility (Morere 1971); the beetle Tenebrio molitor was similarly thought to have no requirement but has subsequently been shown to grow faster in synthetic basal diet if supplemented with linoleic acid (Davis and Sosulski 1973). One may conjecture that such variable results reflect differing levels of pre-experimental reserves of fatty acid in the eggs used by different workers, or that in the earlier studies using casein-based diets some residual milk-derived essential fatty acids were still present as contaminants in the casein. Be that as it may, because the literature contains well-authenticated cases in which fatty acid deficiency symptoms appeared only after a second generation of deprivation was imposed (Earle et al. 1967), negative findings from single generation studies, in early work especially, are best viewed cautiously and tentatively.

Another basis for suspecting a general requirement for dietary polyunsaturates among insects invokes two premises: that animals in general require specific polyunsaturated fatty acids in structural membrane phospholipids, as considered to be the case for vertebrates; and that insects, like vertebrates, would be unable to biosynthesize such polyunsaturated fatty acids. Were these premises valid, it would follow that all insects must need dietary essential fatty acid unless it were provided by symbiotes. With respect to a need for polyunsaturated fatty acids in tissue membrane structures, linoleic and linolenic acids generally preponderate in the phospholipids of insects, and longchain polyunsaturates have recently been found in especially high concentration in phospholipids of several species, and especially in membrane-rich nervous tissue (Stanley-

Samuelson 1980, personal communication), presumably being metabolized and sequestered there for some necessary function. As for an inability to biosynthesize fatty acids with more than one double bond, this was, and perhaps still is, the generally accepted view (e.g., Downer 1978). However, with the recent publication of very compelling evidence for the substantial synthesis of linoleic acid by certain insects (Blomquist et al. 1982) the generalization becomes fragile, and with it the surmise that all insects probably require dietary polyunsaturated fatty acid. This question, so basic to a comprehension of insect fatty acid nutrition, and indeed to the orthodoxies of animal fatty acid metabolism in general, is taken up in greater detail in a subsequent section.

Regardless of the manifestations of fatty acid deficiency in any particular insect, it was at first believed that the requirement could be met interchangeably by either linoleic or linolenic acids. However, this subsequently proved not always to be so among Lepidoptera, the group for which the greatest number of species have been studied and in which the deficiency is expressed most obviously. The earliest studies, of flour moths, already hinted at a greater potency for linolenic acid than for linoleic acid, and a similar difference in potency was clearly detected for the pink bollworm, Pectinophthora gossypiella (Vanderzant et al. 1957). Later, linolenic acid specifically proved essential for complete development of the tea tortrix, Adoxophyes orana (Tamaki 1961), and shortly thereafter the cabbage moth, Trichoplusia ni, was found to require linolenic acid for proper adult emergence, linoleic acid being completely ineffective in this regard although it was required in addition to linolenic acid for optimal larval growth (Chippendale et al. 1964). For several other species of Lepidoptera the pupal/adult ecdysis has now been found linolenate-dependent. A recent summary for 18 species studied in this respect listed 3 of them as requiring no fatty acid at all, 6 utilized linoleic and linolenic acids equally well, 3 could utilize either fatty acid but with linolenic acid the more potent, 5 required linolenic acid specifically, and one apparently required both linoleic and linolenic acids for normal adult emergence (Dadd 1981). It should be emphasized that the foregoing summary refers to the requirement for normal adult emergence only, since several of these species required linoleic acid additionally to optimize larval growth rate, and in some cases even oleic acid had a beneficial effect on the latter parameter (Nakasone and Ito 1966; Turunen 1974).

Further complexity was introduced into insect essential fatty acid nutrition by the recent finding that several species of the dipteran family Culicidae, the mosquitoes, cannot satisfy their requirement for essential fatty acid with linolenic or linoleic acids but must obtain arachidonic acid or certain related longchain polyunsaturates from their diet. This is especially surprising because other insects in which fatty acid deficiency could be remitted by linolenic or linoleic acids failed to respond to arachidonic acid or other long-chain polyunsaturates when these were tested as substitutes for the C18 essential fatty acids (Dadd 1981). This disparity will be taken up later after out-lining what is now known of the mosquito requirement.

ESSENTIAL FATTY ACIDS FOR MOSQUITOES

Mosquito larval nutrition has a considerable liter-ature extending back over 4 decades. Most studies dealt with Aedes aegypti, and none detected any need for die-tary fatty acid over one cycle of larval growth to the development of sometimes vigorous adults, a finding in conformity with the general lack of evidence for an essential fatty acid requirement in all Diptera studied hitherto (Friend 1968, Sang 1978). It will be seen later that A. aegypti is the least susceptible to fatty acid deficiency of the several mosquito species now found to have this requirement, and this, in retrospect, would largely account for the early successes with diets lack-ing fatty acid. It is also worth noting that a cryptic source of essential fatty acid might have been present in some of the earlier diets; those based on casein may have contained traces of active fatty acids as impurities carried over from milk fat, and the ovo-lecithin and cephalin necessary for good growth in some studies (Gol-berg and DeMeillon 1948; Singh and Brown 1957) would certainly have provided essential fatty acid.

When routine nutritional investigation of Culex pipiens first became possible with the development of completely defined larval diets it was noted that al-though adult production was good, adults emerged in a state of weakness such that they rarely were able to fly and rapidly become entrapped at the medium surface (Dadd and Kleinjan 1976). This collapse of the teneral adult pointed to some unappreciated nutrient deficiency, per-haps of an obscure trace metal or a lipid vitamin or growth factor. Unsuccessful attempts to rectify the failure with mineral supplements to the diet such as the ash of stock-reared mosquitoes or an ultra-trace mineral mixture seemed to rule out a need for additional inorgan-ic nutrients. Supplementation of the diet with fat-soluble vitamins and lipid growth factors known from vertebrate nutrition, singly and in various combinations, also was uniformly unsuccessful. This phase of the work included initial tests of linolenic and linoleic acids, since these were known to be widely required by other insects, but without beneficial effect.

Flying adult mosquitoes were first obtained by supplementing basal diet with mammalian serum lipoprotein fractions. The rationale for their use was that such lipoproteins could be expected to supply the spectrum of vertebrate lipid growth factors in a water-soluble and probably better absorbable form, the suspicion being that prior tests of pure lipid growth factors might have failed through inadequate dispersion or absorption from aqueous dietary media. When flight-active lipoproteins were fractionated, activity was found entirely in the lipid moiety, and since this consisted largely of phospholipid (ignoring sterol, an essential insect nutrient routinely provided in all mosquito diets), a range of phospholipids was tested.

The clearcut result was that vegetable lecithins and synthetic dipalmitoyl cephalin and lecithins were inactive, whereas animal-derived lecithins and cephalin supported the emergence of flying adults (Dadd and Kleinjan 1978). This finding pointed to a flight factor associated with the particular fatty acid constituents of animal phospholipids, since all other components of phospholipid molecules would be variously present in both active and inactive kinds. The literature indicated that plant phospholipids would contain linoleic and/or linolenic acids as their only polyunsaturates, and the synthetic phospholipids contained only palmitic acid; in contrast, animal phospholipids would be expected to contain substantial proportions of arachidonic acid and similar long-chain polyunsaturates.

When put to the test, pure arachidonic acid was indeed a potent flight-inducing agent, effective even if supplied only late in development, whereas linoleic acid was without substantial flight-inducing activity (Dadd and Kleinjan 1979). In passing we may note that the effectiveness of arachidonic acid when administered only in the final larval instar is very reminiscent of the late requirement for linolenic acid demonstrated in certain Lepidoptera.

The effects on teneral adults of many pure fatty acids (Dadd 1980, 1981) are summarized in Table 1. All saturated or monoenoic fatty acids tested were without beneficial effect on emergent adults. Six polyunsaturates, comprising members of both the $\omega 6$ and $\omega 3$ families, and with three to six double bonds in the carbon chain, induced predominantly flying adults. A further 6 polyunsaturates, both $\omega 6$ and $\omega 3$ and including di- and tri-enoic members, were considered semi-active because of their ability to support emergence of many adults that could stand on the surface of the medium or hop from it, occasionally with weak flight.

The active fatty acids have in common a particular structural feature, diagrammed in Fig. 5 and in more detail elsewhere (Dadd 1980, 1981). This feature is the possession of 3 double bonds in divinyl methane rhythm counting carboxyl-wards from the 6th carbon from the methyl termination, that is to say, from the ω6 carbon. Additional double bonds to either side of this grouping do not negate activity (and hence the fact that there are both ω6 and ω3 fatty acids active for this mosquito), but a tendency was noted for the flight index to be lower and long-term survival of adults to be poorer for the less unsaturated fatty acids (Dadd 1980). The semi-active fatty acids likewise share a common structure, comprising the two most methyl-wards of the fully active triplet of double bonds; an additional double bond methyl-wards, in the ω3 acids, improves the degree of semi-activity, which

TABLE 1. Fatty acids tested for flight inducing activity with <u>Culex</u> <u>pipiens</u>. Inactive: adults mostly trapped at surface of medium on emergence, with only occasional standers and hoppers. Semi-active: adults commonly stand or hop and occasionally fly weakly. Active: adults predominantly fly.

Fatty acid		Flight Index range	Rating
16:0	palmitic	0-4	inactive
18:0	stearic	0-3	inactive
20:0	arachidic	0-7	inactive
22:0	behenic	0-3	inactive
18:1ω9	oleic	0-4	inactive
18:1ω12	petroselenic	0-3	inactive
20:1ω9		0-1	inactive
22:1ω9	erucic	0-1	inactive
18:2ω6	linoleic	0-14	semi-active
20:2ω6		4-9	semi-active
22:2ω6		8-10	semi-active
18:3ω3	α-linolenic	6-23	semi-active
20:3ω3		0-39	semi-active
22:3ω3		12-26	semi-active
18:3ω6	γ-linolenic	33-79	active
20:3ω6	homo-γ-linolenic	23-68	active
20:4ω6	arachidonic	69-100	active
22:4ω6		84-93	active
22:4ω6		84-93	active
20:5ω3	eicosapentaenoic	70-83	active
22:6ω3	docosahexaenoic	71-94	active

is also noticeably better for the longer-chained C20 and C22 fatty acids than for the C18 members.

At this juncture it is helpful to emphasize the novel features of this C. pipiens fatty acid requirement. First, no species of Diptera previously studied had been found with an essential fatty acid requirement of any sort. Next, neither linoleic nor linolenic acid was fully effective for C. pipiens, though one or the other is satisfactory for all other insects shown to require essential fatty acid. And lastly, the group of fully active fatty acids includes those which are the physiologically important entities for vertebrates, whether obtained as such in the food or metabolized from shorter and less unsaturated dietary polyunsaturates.

A point of interest concerning quantitative aspects of the mosquito requirement can be made in relation to both other insects and vertebrates. The C. pipiens requirement seems very low, the minimum concentration of arachidonic acid needed in the medium for optimal effect being only about 0.0002%, or approximately 0.005% of the total nutrient solids contained in the medium. This may be contrasted with minimal requirements of the order of 0.1% of dietary solids for other insects requiring dietary linoleic or linolenic acid, an optimal requirement of the same order as for vertebrates. The mosquito minimal requirement is comparable in magnitude to that for certain water-soluble vitamins needed in the synthetic medium (Kleinjan and Dadd 1977), and this suggests that the physiological functions underlying this nutritional need are vitamin- or hormone-like, or, if structural, confined to a quantitatively minor type of tissue, such as lipid membranes of muscle, glandular or nervous cells.

Though the foregoing account applies in detail only to C. pipiens, an essential fatty acid requirement that can be satisfied by arachidonic acid has now been demonstrated for several mosquito species. Some years ago it was noted that 10 species from 3 genera could be reared from egg to adult with varying degrees of success on the basal medium developed for C. pipiens (Dadd et al. 1977); in all cases except for A. aegypti, any adults obtained were, like C. pipiens, unable to fly. Several of these and other species have now been shown to produce appreciable numbers of flying teneral adults and sometimes to grow faster as larvae if the dietary medium contains arachidonic acid, as the data presented in Table 2 illustrate.

For each species dealt with in Table 2, three parameters are listed. The developmental index (defined and discussed by Dadd and Kleinjan (1976) provides a combined measure of rate of development and overall amount of development achieved before death by 1st instar larvae set out in the medium; the percentage of adults gives the proportion of 1st instar larvae to become adult, whether trapped at emergence or able to fly; and the flight index (defined by Dadd and Kleinjan 1978) indicates the proportion of teneral adults that can fly. The several species are arranged in three groups, listed in order of increasing apparency of symptoms of arachidonic acid deficiency. Heading the list is A. aegypti, which develops equally well to the adult stage whether or not arachidonic acid is in the diet, and which produces substantial numbers of flying adults even without dietary arachidonic acid, though in the experiment illustrated many more fliers were obtained with the fatty acid present. Teneral adults of all other species failed to fly without arachidonic acid in the diet. They are conveniently considered under two categories: those whose larval development was apparently unaffected by arachidonic acid, so that deficiency became manifest only with the collapse of emergent adults; and those whose larval development was retarded in the absence of arachidonic

TABLE 2. Developmental parameters and flight indices for 7 species of mosquito reared on synthetic diet with and without arachidonic acid at concentrations of 0.4 mg per 100 ml of dietary medium.

Species	Arachidonic acid?	Developmental index	Percent adults	Flight index
Aedes aegypti	without	4.15	95	32
	with	4.10	100	77
Culex pipiens	without	4.90	95	3
	with	4.91	95	95
Aedes sierrensis	without	2.63	93	0
	with	2.76	100	59
Culiseta incidens	without	3.29	71	0
	with	3.88	87	43
Culex tarsalis	without	1.49	32	0
	with	2.34	62	71
Anopheles stephensi	without	1.27	15	0
	with	1.84	50	41
Culiseta inornata	without	1.12	0	-
	with	2.01	41	61

acid and survival during pre-adult development was substantially reduced. C. pipiens exemplifies the category
in which virtually no adverse effect on larval development or survival attends deficiency; for this species,
and also for Culiseta incidens and Aedes sierrensis, the
deficiency becomes apparent in a clearcut way only at
the time of adult emergence. The third category, comprising Culex tarsalis, Culiseta inornata and Anopheles
stephensi, shows a clear larval dependence upon the
essential fatty acid, most marked with C. inornata, nearly all of which died before pupation without arachidonic
acid (Dadd et al. 1981).

Since larval development is virtually unaffected by
deficiency in species such as C. pipiens the impression
could be gained that arachidonic acid is essential
specifically in connection with some aspect of wing
development or flight physiology, especially as it was
possible to avert adult dysfunction in C. pipiens by
supplying fatty acid during the final larval instar only
(Dadd and Kleinjan 1979). However, the fact that several
species show defective larval development and heavy prepupal mortality without arachidonic acid indicates a
probable functional requirement for essential fatty acid
throughout development, though perhaps reaching a cres-
cendo of demand for some especially critical functions
associated with metamorphosis. Most likely the differing stages of development·at which symptoms of deficiency
appear reflect differing reserve levels of essential fatty acid in the eggs from which experimental larvae were
obtained. Of those species for which only the single
experiment documented in Table 2 was carried out one
cannot say whether this is a consistent species characteristic or a chance happening related to the particular
batch of eggs used in the experiment. For A. aegypti,
C. pipiens, C. tarsalis, C. incidens and A. stephensi
the effect, or lack of it, on larval development was
consistent over several experiments and thus appears to
be a species characteristic.

Assuming the differences discussed above do indeed
indicate species differences in egg reserves of essential fatty acids, then A. aegypti evidently has sufficient reserves to carry most individuals through to
viable adult existence, and it would require continua-
tion of fatty acid deprivation into adult reproductive
performance to firmly establish fatty acid essentiality,
as earlier noted for certain cockroaches and beetles.
This has not been done in experiments using the pure
fatty acid, but it has recently been shown that A.
aegypti larvae reared on synthetic diet supplemented
with 0.002% of animal lecithin, a source of arachidonic
acid, produce females of greatly increased longevity and
fecundity than do larvae reared with unsupplemented basal

diet (Sneller and Dadd 1981).

For most of these mosquito species it has not yet been determined which fatty acids other than arachidonic acid are or are not effective in remitting the essential deficiency. Nevertheless, for C. tarsalis the pattern of effective and inactive fatty acids is essentially the same as for C. pipiens; for A. stephensi neither linoleic nor Δ11,14,17-eicosatrienoic acids are able to induce flying in teneral adults (Dadd unpublished data). On present evidence it therefore seems probable that a pattern of essential fatty acid utilizability similar to that worked out for C. pipiens will hold for the Culicidae as a whole.

CAN POLYUNSATURATED FATTY ACIDS BE BIOSYNTHESIZED BY ANY INSECT?

At the very genesis of interest in essential fatty acids for insects this question received an affirmative answer from an extension of Fraenkel and Blewetts' work with the flour moth. Having demonstrated for Ephestia a clear-cut response, in terms of faulty adult emergence, to a deficiency of linoleic or linolenic acids, it was realized that the presence of these essential fatty acids in tissues of any other insect could be bioassayed by incorporating lipid extracted from the insect of interest into fatty acid-free diet for Ephestia larvae and determining whether or not the resultant moths emerged perfectly. If emergences were normal, then the extracted insect must have contained fatty acids essential for Ephestia. Further, if the test insect had been reared from the egg on a diet devoid of fatty acid, then it must itself have biosynthesized the ephestia-essential fatty acids detected by the bioassay (perhaps with the assistance of symbiotes, if such were present). Applying this bioassay to the mealworm, Tenebrio molitor, Fraenkel and Blewett (1947) concluded that this beetle, considered to have no dietary essential fatty acid requirement on the basis of prior nutritional studies, was able to biosynthesize linoleic or linolenic acid.

A similar bioassay was recently used to demonstrate that tissues of various insects contained arachidonic or related fatty acids essential for the mosquito Culex pipiens (Dadd 1981; Stanley-Samuelson and Dadd 1981). However, since the advent of gas-chromatography in the 1960s, which afforded fine quantitative resolution of fatty acid mixtures extracted from tissues, the possibility of biosynthesis has usually been examined by direct analytical methods coupled with isotope labelling. If an insect can be reared from the egg on a defined diet devoid of fatty acid, chromatographic resolution of the fatty acids extracted from its tissues reveals directly those that can be synthesized de novo. For example, such a study led to the conclusion that the beetle <u>Dermestes maculatus</u>, though apparently requiring no dietary fatty acid for growth from egg to adult, was unable to synthesize linoleic and linolenic acids (Cohen 1974). Alternatively, and especially if complete rearing on defined diet is impossible, the levels of radioactivity associated with individual fatty acids separated by gas chromatography from tissue fatty acids of insects fed or injected with isotope-tagged precursors of fatty acid biosynthesis such as glucose or acetate may be determined after a suitable lag time for metabolism; the detection of radioactivity in a particular fatty acid indicates some biosynthesis of it, while the absence of tracer suggests none. As examples, <u>Drosophila melanogaster</u> and <u>Hyalophora cecropia</u> were thus shown unable to biosynthesize either linoleic or linolenic acids from isotopic acetate, nor could these polyunsaturated fatty acids be derived from radiolabelled C16 and C18 saturated or monoenoic fatty acids (Keith 1967, Stephen and Gilbert 1969).

The fatty acid metabolism of many insect species has now been examined using one or other of the foregoing methodologies, with the general finding that fatty acids having more than one double bond are not, or are only marginally, synthesized. Reviewing all evidence then available, Gilbert (1967) concluded that although some insects clearly could not synthesize polyunsaturates, others might be able to synthesize all the fatty acids they required physiologically, including di- and tri-enoic acids. However, a few years later Gilbert and O'Connor (1970) concluded that no arthropod has been shown to synthesize C18 polyunsaturated fatty acids and Fast (1970) categorically stated "Polyunsaturated fatty acids are not synthesized by insects", a view subsequently supported in the main by Downer (1978), and one which nicely complemented the increasing evidence for a nutritional polyunsaturated fatty acid requirement in most insects (Dadd 1973, 1977).

The uncertainty lurking behind these reviews reflects the fact that from the beginning of metabolic studies of fatty acid biosynthesis cases were reported in which small but significant amounts of radiolabel appeared in linoleic or linolenic acids, raising the possibility that a real, if marginal biosynthesis of polyunsaturate might be occurring. For insects known to harbor gut or intracellular symbiotes, such as the aphid <u>Myzus persicae</u>, the apparent low-level incorporation of radioactivity into C18 di- and tri-enoic fatty acids (Strong 1963) could be discounted as evidence for the insect's biosynthetic capability because of possible symbiote propensities for polyunsaturate manufacture. Similar results with cockroaches (Louloudes et al. 1961) could likewise be discounted because of the complex symbiotic associations of these insects, especially when isotope incorporation into linoleic acid was found to be lower in roaches reared aseptically and thus lacking gut symbiotes, though still with intracellular fat-body symbiotes (Bade 1964). A further reason for ignoring relatively minor levels of isotope incorporation into linoleic or linolenic acids was to regard them as artefactual, resulting from incomplete chromatographic separation of, e.g., oleic and linoleic acids or linolenic and arachidic acids, the monoenoic and saturated members of these pairs being known to be synthesizable from acetate. These artefactual problems were critically discussed by Stephens and Gilbert (1969) and doubtless led to the changing interpretations noted above.

Nevertheless, brushing aside instances of possible low level contamination of polyunsaturate peaks or obfuscation by symbiotes, there remained awkward reports in which linoleic or linolenic labelling was really rather high, approaching levels for the commonly biosynthesized fatty acids such as oleic, stearic, palmitic and palmitoleic acids. This was so for studies of fatty acid biosynthesis by whole spider mites (Walling et al. 1968). Also, whole body fatty acid analysis of the cricket, Acheta domesticus, reared from the egg on casein-based meridic diet devoid of added fatty acid, revealed exceptionally high proportions, up to 20%, of linoleic acid, implying its de novo biosynthesis at rates comparable with those for the saturated and monoene fatty acids whose biosynthesis was well authenticated in all metabolic studies (Meikle and McFarlane 1965).

Most recently, very careful metabolic and biochemical studies have shown large-scale synthesis of linoleic acid from isotopic acetate in a cockroach, Periplaneta americana, a termite, Zootermopsis angusticollis, and also in the cricket Acheta domestica, the latter confirming the findings from simple fatty acid analysis noted above; control studies on a different roach, Blattella germanica, an aphid, Acyrtosiphon pisum, and the common house fly, Musca domestica, gave the more usually expected result of no linoleate biosynthesis (Blomquist et al., in press). These results are curious in that biosynthesis or lack of it for the 6 species concerned seems unconnected with either taxonomic relationships (the two cockroaches) or the presence or absence of known symbiotic associations (the roaches, the termite and the aphid all are symbiote-dependent). These results appear to undermine the assumption that insects, and indeed, animals in general, cannot biosynthesize polyunsaturated fatty acids, and this must substantially modify speculations about the place of essential fatty acids in insect nutrition.

COMPARISONS AND SPECULATIONS

I recently put forward speculations intended to knit what was known of insect and vertebrate essential fatty acid requirements and physiological functions into a unified scheme (Dadd 1981). With the passage of a year, additional evidence supports some of these conjectures but has negated others. Table 3 sets out for comparison some pertinent generalizations from the information then available on vertebrate and insect essential fatty acids. In this summary, question marks indicate propositions for which information was sparse, lacking entirely, or which I then thought might be equivocal on technical grounds; in parentheses I indicate changes consequent upon evidence which has accrued since the review was written.

The new information on mosquito requirements for longchain polyunsaturates suggests a link between the physiological needs of this particular group of insects and those of vertebrates, for some or other of which the same polyunsaturates are not only dietarily adequate but are also the fatty acid entities of central physiological importance. However, mosquitoes differ from the generality of vertebrates in being unable to avert fatty acid deficiency with dietary linoleic or linolenic acids, suggesting a lack of the unsaturated fatty acid elongases and desaturases that are generally available to vertebrates and allow them to convert C18 polyunsaturates of food into the physiologically necessary higher polyunsaturates. A vertebrate exception, the cat, lacks appropriate desaturases (Rivers et al. 1975) and so, like mosquitoes, must eat long-chain polyunsaturates. Probably some $\omega 3$-requiring fish also have an impaired ability to metabolize linolenic acid to higher polyunsaturates since they require dietary eicosapentaenoic or docosahexaenoic acids (Yone and Fujii 1975).

Turning to insects other than mosquitoes, they share with vertebrates the ability to utilize linoleic or linolenic acids as their dietary essential fatty acid, but in contrast to vertebrates and mosquitoes, all tests of the utility of dietary arachidonic or docosahexaenoic acids have found these higher polyunsaturates inadequate. In apparent harmony with this, the mass of analytical data on fatty acid composition of insect tissue lipids has hitherto been notable for an almost complete absence of records of the higher polyunsaturates commonly found in vertebrate fats, which could be taken to imply that other insects as well as mosquitoes lacked desaturating and elongating enzymes. Hence one might conclude that for insects requiring C18 fatty acids in the diet, these are also the effective physiological fatty acids, in marked contrast to both the vertebrate and mosquito situations. However, higher polyunsaturates are also

TABLE 3. Comparison of some generalizations respecting polyunsaturated fatty acid dietary requirements and detection in tissues for vertebrates and insects.

| | Dietary essential fatty acids | | Fatty acid metabolites in tissues | |
	linoleic or linolenic effective	C20 or C22 polyunsaturates effective	C20/22 acids commonly detected	Prostaglandins detected
Most vertebrates	yes	yes	yes	yes
Cat	no	yes	yes	yes
Mosquito	no	yes	no? (recently yes)	no(?)
Other insects	yes	no? (possibly)	no? (recently yes)	2 spp. (now 5)

absent from most analytical records for mosquito material, though mosquitoes require them dietarily, and bio-assay evidence indicated their existence in mosquito tissues; in Table 3 I therefore questioned the validity of negative findings with respect to tissue polyunsaturates for both mosquitoes and other insects.

Lastly, a well defined function of essential fatty acid in vertebrates is to provide the C20 precursors for biosynthesis of prostaglandins, probably universally present in this phylum. In contrast, prostaglandins were unknown in insects until 5 years ago, and as recently as 2 years ago had been detected only in reproductive tissues of 2 species, both crickets. Thus, at the time of my earlier review, this also was an equivocal area.

From the comparisons of Table 3 it will be seen that disparities between vertebrates, mosquitoes, and other insects would be reduced if any or all the following generalizations proved invalid: 1) Insect tissues generally lack highly unsaturated long-chain polyunsaturates; 2) Insects requiring dietary linoleic or linolenic acids cannot utilize higher polyunsaturates in their place; and 3), Prostaglandins or related metabolites occur only as occasional curiosities in insects. Previously I argued that all these propositions must now be considered tentative since the evidence for their generality had in some cases already been seriously broached. It was my speculative view that in ultimate physiological terms the essential fatty acid requirements of all insects and vertebrates would, in broad outline, prove homologous, though with ω6 and ω3 specialists among both taxa. Evidence recently accumulated does indeed further erode propositions 1) and 3), but a recent attempt to obtain further evidence, confirmatory or negatory, bearing on proposition 2) has led to greater complexity rather than clarity for this crucial issue.

Long-chain Polyunsaturates in Insects

Though some pre-chromatographic analyses indicated substantial proportions of arachidonic acid or a similar tetraene in fatty acids of a locust (Fawzi et al. 1961) and a moth (Schmidt and Osman 1962), subsequent gas chromatographic analyses for a multitude of insect species rarely recorded polyunsaturates of carbon chain length greater than 18 (Fast 1964, 1970). Previous discussions of this literature (Dadd and Kleinjan 1979, Dadd 1980) point out that lack of evidence for higher polyunsaturates encompassed numerous studies of mosquito material, which seemed especially surprising once it had become clear that such polyunsaturates were most probably a general dietary requirement for mosquitoes. There were a few exceptions to the general absence of higher poly-

unsaturates in this extensive insect fatty acid literature. Arachidonic acid was detected in the cockroach
Periplaneta americana (Kinsella 1966) and in larvae of
two moths, Bombyx mori (Nakasone and Ito 1967) and Heliothis virescens (Wood and Harlow 1969). Though an earlier
paper reported none (Wright and Oehler 1971), subsequent
work found many longchain polyunsaturates in the hornfly
Haematobia irritans (Mayer and Bridges 1975), ascribed
to the use of a column more appropriate to longchain
polyunsaturates. The more recent work on mosquitoes
found small amounts of several unidentified long-chain
polyunsaturates (Schaefer and Washino 1969, 1970), and
cultured mosquito cell lines also contained arachidonic
acid and other polyunsaturates (Jenkin et al. 1975,
1976). Most interestingly, retinal phospholipids from 3
insect species, a lepidopteran, a dipteran, and a neurop-
teran, contain appreciable arachidonic acid and major
amounts of eicosapentaenoic acid (Zinkler 1975).

Using a bioassay similar to that discussed above in
connection with the flour moth, lipid extracts from
mosquitoes and several other taxonomically diverse
insects were shown to contain pipiens-essential fatty
acids (Dadd 1981). Subsequently, arachidonic, gamma-
linolenic, and other long-chain polyunsaturated fatty
acids were demonstrated by gas chromatography in extracts
of both natural- and synthetic diet-reared C. pipiens,
being especially prominent in phospholipid-rich fractions
separated from such extracts by thin layer chromatography; moreover, the proportions of arachidonic acid
detected correlated well with the concentrations of pure
arachidonic acid provided in synthetic diets in which
larvae of extracted mosquitoes had been reared (Stanley-
Samuelson and Dadd, 1981).

Similar gas chromatographic studies have now demonstrated C20 or C22 polyunsaturates in all of a dozen or
so insect species of diverse taxa and dietetic habit
(Stanley-Samuelson 1980 and personal communications). In
crude lipid extracts of whole insects the long-chain
polyunsaturates generally constitute no more than 2% of
the fatty acid total, but may be present in 5-10 times
this proportion in phospholipid fractions separated from
the crude extracts, being virtually absent from the corresponding triglycerides; they also were found in high
concentrations in extracts made from particular tissues
such as those of the nervous system. Of particular
interest is the presence of long-chain polyunsaturates
in phytophagous species such as bees and lepidopterous
larvae; such insects presumably were restricted in their
dietary polyunsaturated fatty acid intake to linoleic
and linolenic acids, and so most likely biosynthesized
the characteristically animal polyunsaturates detected
in their tissues from the C18 plant fatty acids of their

food. The observation of greatly enhanced proportions of higher polyunsaturates in phospholipid, and especially in specific tissues, is of great significance in relation to the idea that they are functional rather than adventitious. Even in an insect such as the cockroach, which might accumulate polyunsaturates adventitiously from an omnivorous diet, sequestration into particular tissues implies special functional needs for polyunsaturates in those tissues.

These extensive positive data covering many species of diverse taxa suggest strongly that most insects would now be found to possess polyunsaturates if appropriate tissues and/or lipid classes were examined with instrument sensitivity maximized to detect long-chain polyunsaturates. That they were missed or not reported in nearly all past studies doubtless reflects their generally low or trace proportionalities, especially in whole body extracts where the preponderant triglyceride fatty acids from the fat body would especially dilute them. When dealing with materials that occur as minor components, what is detected is to a considerable extent what is expected, and this doubtless also contributed to the past negative record. The history of cricket fatty acid analyses is instructive in this context. Early studies of Acheta domesticus found no C20 polyunsaturates (Meikle and McFarlane 1965), and their absence was specifically noted for Gryllus bimaculatus (Fast 1967). Then prostaglandins and prostaglandin synthetase were detected in A. domesticus, the first record for an insect (Destephano and Brady 1977), implying the presence of arachidonic acid as a necessary precursor. Most recently, and presumably with an incentive to search for it critically, arachidonic acid has been found in tissues of this cricket (Worthington et al. 1981).

Since, as now seems probable, insects generally contain some higher members of the ω3 or ω6 polyunsaturate families, if not obtained directly from their food, they must, as in vertebrates, be derived by elongation and desaturation of parent C18 fatty acids in the food, or, as we may now anticipate in some insects (Blomquist et al. 1982), from biosynthesized linoleic acid. In special cases such as the cat, mosquitoes, and certain fish, tissue polyunsaturates must be acquired pre-formed from the diet, consequent upon a lack of appropriate desaturases in the cat and doubtless so also for the mosquitoes and fish.

Prostaglandins in Insects

Essential fatty acids are thought to subserve two main physiological functions in vertebrates: as components of specific phospholipids they seem important to the integrity of cellular lipid membranes and membrane-associated enzyme activities; and they provide the C20 fatty acid precursors for the hormone-like prostaglandins needed for localized metabolic regulation in many tissues. By analogy with sterol utilization in insects we might think of these as "structural" and "metabolic" functions respectively. To clarify this analogy, all insects require dietary sterol, a major physiological function of which is structural, since sterols, like phospholipids, are necessary components of cellular membranes; another use of dietary sterol, the so-called metabolic function, is to provide precursors for the important group of ecdysone moulting hormones. The range of sterols able to subserve the structural function is generally broader than for the metabolic function, and from this arises the phenomenon of "sterol sparing". For most insects, dietary cholesterol is sufficient alone for both structural and metabolic functions. However, it is often the case that with only a small proportion of the total required sterol provided as cholesterol and presumed to underwrite the metabolic function, some other sterol, which on its own would be inadequate for all functions, is able to take over the less fastidious demands of the structural requirement. Such other sterols are then termed sparing sterols (Dadd 1973). For example, cholestanol can spare at least 95% of the dietary cholesterol requirement of <u>Drosophila melanogaster</u> (Kircher and Gray 1978).

Since semi-active and fully-active fatty acids could be distinguished for <u>C</u>. <u>pipiens</u>, a concept of essential fatty acid sparing, analogous to sterol sparing, seemed not unreasonable; it could be imagined that both fully-active and semi-active fatty acids might satisfy a structural role, but with only the fully-active ones additionally able to provide precursors for some essential vitamin- or hormone-like entity. Figure 3 illustrates an experiment to probe this hypothesis. Arachidonic acid in a diminishing concentration series was compared with the same arachidonic concentration series combined with a fixed amount of C20:3ω3, a semi-active fatty acid. For the serial concentrations of arachidonic acid on its own, flight indices are progressively lower as concentrations fall below 0.05 mg/100 ml diet, but with the semi-active fatty acid also present, high flight indices are maintained down to an arachidonic acid concentration of 0.01 mg/100 ml. The semi-active fatty acid thus appears to spare some fraction of the amount of arachidonic acid needed on its own to achieve high flight activity, an indication of two distinct functions for the latter.

Because of the well known relationship between arachidonic acid and prostaglandins, the latter were obvious candidate derivatives for a putative metabolic function.

Straight replacement of arachidonic acid in diets by various prostaglandins proved uniformly ineffective, as
would be expected if arachidonic acid were bifunctional
and provided structural fatty acid as well as being a
prostaglandin precursor. I therefore tried the approach
of using the semi-active fatty acids C20:3ω3 or C22:3ω3
in combination with a prostaglandin (PGE2 or PGA2) on the
hypothesis that the semi-active "sparing" fatty acid
might fill the structural role while the prostaglandin
might directly replace the postulated metabolic function
of arachidonic acid. In none of these experiments were
such fatty acid/prostaglandin combinations able to induce
a level of flight activity comparable to that induced by
a fully-active fatty acid alone. In some, supplementation with PGA$_2$ seemed to increase by about 50% the
flight index obtained with the semi-active fatty acid
alone, but just as often no effect was observed, and no
effect was ever observed using PGE$_2$. Thus, this line
of enquiry is currently indeterminate, perhaps not sur-
prisingly, in view of the failure in most vertebrate
studies to demonstrate any essential fatty acid replace-

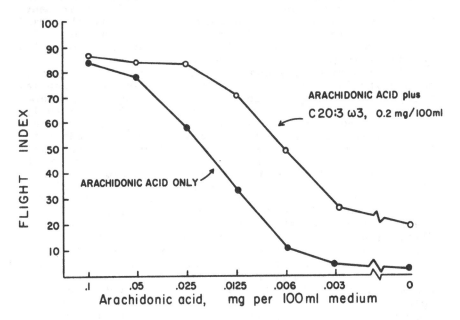

FIGURE 3. Plots of the flight index for decreasing concentrations of a pipiens-essential fatty acid, arachidonic acid, with or without a constant level of the
semi-essential eicosatrienoic acid.

ment abilities for prostaglandins (Lands et al. 1977). It is perhaps expecting too much to hope that such labile materials would remain active in dietary media long enough to be ingested or absorbed in amounts sufficient to show much effect; moreover, of the many active prostaglandins known from mammallian physiology only two of the commoner ones have so far been tested with mosquitoes, and these might well be inappropriate.

In spite of the failure to demonstrate prostaglandin action directly, I have obtained circumstantial evidence for a prostaglandinogenic function for arachidonic acid in C. pipiens by the use of prostaglandin synthetase inhibitors. When these are included in rearing media containing slightly suboptimal levels of arachidonic acid, flight indices are reduced, a reduction that can be counteracted by increasing the dietary concentration of arachidonic acid. This type of effect has been found consistently using phenylbutazone or indomethacin, though only erratically and less clearly with aspirin. Figure 4 graphically presents data from such experiments with phenylbutazone. With a constant phenylbutazone concentration versus three doubling concentrations of arachidonic acid, reductions of flight indices by 74%, 44%, and 34% respectively compared with controls lacking inhibitor were obtained. An essentially similar pattern resulted for the experiment with indomethacin, except that at the highest arachidonic acid concentration, inhibition was virtually balanced out.

These experiments indicate that the flight inducing effect of arachidonic acid is inhibited by some prostaglandin synthetase inhibitors. This inhibition seems to be progressively reduced the higher the arachidonic acid concentration, and thus cannot be ascribed to some general toxic effect of the inhibitors. Since the inhibiting substances are known prostaglandin synthetase inhibitors in mammalian work, the most parsimonious explanation for these findings is to postulate that arachidonic acid is acted on by prostaglandin synthetases to form prostaglandins which in some way are necessary for the emergence of strong, flight-worthy adult mosquitoes.

While these experiments by no means yet prove that prostaglandinogenesis is an important aspect of the mosquito requirement for arachidonic acid, they are consistent with the accumulating evidence that prostaglandins may be of general importance in insect function and not just curiosities of cricket reproductive behavior. Various prostaglandins have now been detected in Bombyx mori, a lepidopteran (Setty and Ramaiah 1979, 1980), Musca domestica, a dipteran (Wakayama et al. 1980), and in a termite (I. Kubo, personal communication), in all cases associated with reproductive tissues and sometimes with a demonstrated reproductive function, as originally found with crickets (Destephano and Brady 1977; Loher 1979; Loher et al. 1981). Furthermore, prostaglandins have now been detected in non-reproductive cricket tissues (Loher and Kubo personal comm.), suggesting possibly wider functionalities than just reproduction. Because of the wide taxonomic distribution of the 6 species in which the occurrence of prostaglandin has so far been determined directly or indicated by prostaglandin synthetase inhibitor action, one may well expect the burgeoning new field of insect prostaglandin studies to embrace insects in general, a view that derives reinforcement from the recent evidence for the widespread occurrence of at least traces of arachidonic acid in insects of diverse taxa.

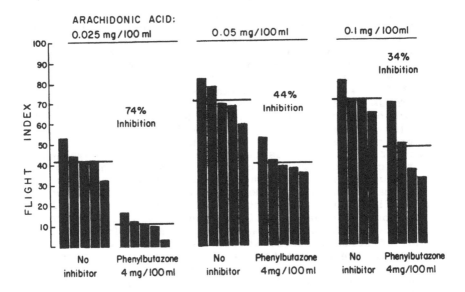

FIGURE 4. Flight indices for 3 concentrations of dietary arachidonic acid with or without fixed dietary concentration of phenylbutazone. Bars give replicate values and horizontal lines indicate means for each treatment. Inhibition is calculated as percent reduction of means with phenylbutazone for treatment pairs at each arachidonic acid concentration.

Can Higher Polyunsaturates Replace Dietary C18 Polyunsaturates?

The higher ω6 polyunsaturates are dietarily interchangeable with linoleic acid for warm-blooded vertebrates, as would be expected given that arachidonate is the physiologically crucial fatty acid derived in their

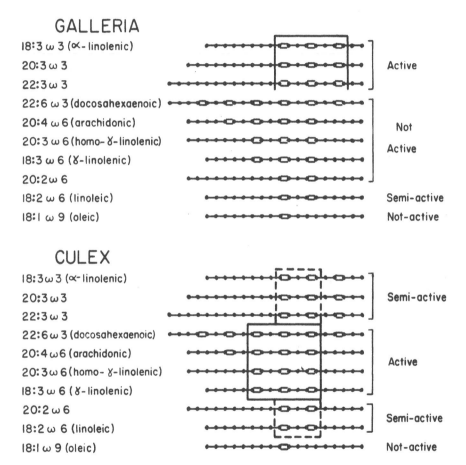

FIGURE 5. Diagrams of fatty acid carbon chains showing structural features of active and semi-active essential fatty acids for <u>Galleria mellonella</u> and <u>Culex pipiens</u>. The active double bond structure is boxed by solid lines and for <u>Culex</u> the semi-active structure is boxed by dashed lines. It is not clear what features gives rise to the semi-activity of linoleic acid for <u>Galleria</u>.

tissues from linoleic acid or other ω6 fatty acids in the diet. We have seen that arachidonic acid and other long-chain polyunsaturates are frequently, perhaps generally, present in insect tissues, presumably metabolized from dietary (or biosynthesized) linoleate or linolenate when not obtained directly from food. Were higher polyunsaturates the physiologically important entities for insects that they are for vertebrates, then one might expect them also to be dietarily adequate for insects, just as for vertebrates. But in fact past evidence is uniformly negative on this point. Arachidonic acid was tested as a substitute for linoleic or linolenic acids with 2 species of locust and 5 species of moth, and docosahexaenoic acid also for 2 of the moths (Dadd 1961, Fraenkel and Blewett 1946, 1947; Chippendale et al. 1964; Dadd 1964; Rock et al. 1965; Rock 1967): in some cases larval growth was distinctly improved by the C20 and C22 polyunsaturates, but in no case was the characteristic essential fatty acid deficiency symptom of faulty adult emergence averted, as it is in all these insects by either linoleic or linolenic acids.

If these results are accepted at face value the conclusion seems inescapable that at the level of physiological function many insects must differ in some fundamental way from the vertebrates in their essential fatty acid usage with respect to either a structural or metabolic role. If it is the metabolic role that is crucially different, it becomes difficult to accomodate the increasing evidence for prostaglandins in insects: if a structural role, one would have to suppose that many insects needed C18 polyunsaturates specifically in their membrane phospholipids and, moreover, lacked an ability to derive them by retroconversion from higher members of the appropriate fatty acid family. On this view, mosquitoes would seem an anomalous group of insects that had evolved essential fatty acid physiological requirements of a sort quite distinct from those of other insects and in many respects strangely parallel to those of vertebrates.

In previous discussions of this situation (Dadd 1981) I called in question the validity of negative findings with respect to the adequacy of arachidonic and docosahexaenoic acids for those few insects in which the matter had been tested. I noted that all these studies involved larval developmental periods of several weeks, during which it was conceivable that oxidative loss might have reduced polyunsaturates in the diets to inadequate levels and so have led to false negative outcomes, especially bearing in mind the evidence suggesting a critical need for essential fatty acid late in development, at metamorphosis or the adult moult. Judging that the matter would remain equivocal until a suitable linoleic- or linolenic-

requiring insect was studied using precautions to avoid
deterioation of long-chain polyunsaturates, my bias inclined to the expectation that arachidonic or docosahexaenoic acids would prove adequate replacements for the C18
polyunsaturates if retested under such conditions.

I am currently re-examining the fatty acid requirements of the waxmoth, Galleria mellonella, using ascorbyl
palmitate-protected fatty acids (Dadd and Kleinjan 1979b)
provided during the final week of larval development. Two
decades ago this insect was shown to have an essential
fatty acid requirement that could be satisfied by either
linoleic or linolenic acids but not, apparently, by arachidonic acid (Dadd 1964). The results obtained to date
in the new study are summarized in the diagrams of Figure
5, which also includes for comparison the findings for C.
pipiens using the same fatty acids. The important points
to emerge are as follows: 1) Linolenic acid is many times
more potent on a weight basis than linoleic acid with
respect to averting failure or deformity at adult emergence; with 0.05% of linolenic acid in the diet adult
emergences approach 100% normal, whereas with this same
concentration of linoleic acid most adults emerge with
severe wing deformities or worse, more than 0.2% being
required to avert wing malformation; linoleic acid is
therefore classed in Figure 5 as semi-active. 2) Arachidonic and docosahexaenoic acids at a concentration of
0.1% are devoid of activity, with most adults trapped in
the pupal integument or grossly deformed, just as for
treatments with oleic acid or no added fatty acid; thus,
the earlier findings that arachidonic acid could not replace the essential C18 fatty acids received confirma-
tion. 3) All other ω6 polyunsaturates tested, except
linoleic, were as inactive as arachidonic acid. 4) The
ω3, C20 and C22 trienoic analogues of linolenic acid were
fully as effective as linolenic acid itself; but docosahexaenoic acid was totally ineffective; thus, certain
higher polyunsaturates, though not the ones anticipated,
can replace the essential C18 polyunsaturates for the
waxmoth.

It is clear from these results that the waxmoth
should now be grouped with those increasingly numerous
species of Lepidoptera discussed earlier as having a
special need for linolenic acid. It also appears that,
in nutritional terms, this need may not be for linolenate specifically, but rather for a broader group of ω3
fatty acids having a certain pattern of double-bonding,
the precise nature of which the data do not yet allow to
be specified, beyond noting that too many double bonds,
as in docosahexaenoic acid, negate activity. Apparently
carbon chain length over the range 18-22 is not crucial
for nutritional adequacy, which suggests that one or
other of two situations with respect to polyunsaturated

fatty acid metabolism obtain: 1) the physiologically required fatty acid is indeed linolenic acid and the C20 and C22 analogues are simply retroconverted by chain-shortening to linolenic acid after uptake into the tissues; or 2) longer-chained $\omega 3$ fatty acids are physiologically required and are either the active analogues themselves or some more unsaturated members such as $C20:4\omega 3$, $C20:5\omega 3$, $C22:5\omega 3$, etc., for which the tested active analogues would be metabolic intermediates on a pathway from linolenic acid. One may doubt that retroconversion of the active C20 and C22 analogues to linolenic acid is probable, since an analogous conversion of the various $\omega 6$ C20 fatty acids to semi-active linoleic acid appears to be ruled out by their complete nutritional inactivity. With respect to the $\omega 3$ requirement, this would then favor the second alternative; that the nutritional importance of linolenic acid derives from its metabolic conversion to a higher C20 or C22 $\omega 3$ fatty acid as the physiologically required entity.

The preceding comment is restricted to the $\omega 3$ requirement because there is still the question of the semi-active status of linoleic acid to be taken into account. Does linoleate satisfy in an inefficient way the same functional requirement(s) that the $\omega 3$ fatty acids satisfy efficiently? Is there a quite separate and independent functional need for linoleate? Or do $\omega 3$ fatty acids satisfy two separate functions, one of which linoleic acid can spare?

Several previous studies with Lepidoptera found that while only linolenic acid could assure a normal adult emergence, and though in some cases linolenic acid also affected larval growth positively, linoleic acid was additionally necessary for optimal growth (Chippendale et al. 1964; Nakasone and Ito 1966; Turunen 1974). Further, one recent study presented evidence that linoleic and linolenic acids are both essential for normal adult emergence (Sivapalan and Gnanapragasam 1979). The new waxmoth studies do not address this question since larval growth rates were not considered; and though the apparently complete adequacy of the active $\omega 3$ fatty acids for normal adult emergence would seem to exclude linoleic acid at least from this aspect of deficiency, this too is uncertain, since low levels of linoleic acid may have been present in all diets as milk fat-derived impurities in the casein. Nevertheless, the weight of current evidence from many Lepidoptera indicates a requirement for both linoleic and linolenic acids (or in the latter case, a particular sub-group of the $\omega 3$ family of fatty acids). Whether this might also be the case for insects of other orders previously shown to require C18 polyunsaturated fatty acids cannot be judged on evidence now available; a re-examination of, for example, the ability of locusts to utilize a range of $\omega 6$ and $\omega 3$ fatty acids would do much to clarify the situation more generally.

Mosquitoes, however, do not readily fit into such a pattern of double fatty acid requirements. Figure 5 shows that essential fatty acid structure/function relationships for C. pipiens are quite different from those for the waxmoth; fatty acids which fully satisfy the essential requirement of C. pipiens include both $\omega 3$ and $\omega 6$ polyunsaturates, any one of which seems adequate on its own for all essential functions, at least over one generation. In the case of arachidonic acid, more than 10 sequential generations have been maintained axenically on synthetic diet containing it (Dadd and Kleinjan, unpublished), virtually eliminating the possibility that some other fatty acid carried over from egg reserves might also be essential.

There is, nevertheless, an interesting difference with respect to fatty acid metabolism between synthetic diet-reared C. pipiens, with or without arachidonic acid in the diet but with no other fatty acids, and normal stock-reared mosquitoes. In the former, only low or trace amounts of linoleic acid and linolenic acids are present in tissues, whereas these fatty acids form a substantial proportion of stock-reared mosquito fatty acids, presumably acquired from the crude natural diet; and in the synthetic diet-reared mosquitoes palmitoleic acid (C16:1) occurs in double the proportions found in the stock-reared mosquitoes (or recorded in other species reared naturally), constituting about 50% of all fatty acids (Stanley-Samuelson and Dadd 1981). This preponderance of palmitoleic acid is very similar to the increased proportions of both oleic and palmitoleic acids found in tissue lipids of other Diptera and Lepidoptera reared on diets lacking C18 polyunsaturated fatty acids (Keith 1967; Nakasone and Ito 1967; Turunen 1973, 1976), the increase of palmitoleic acid in C. pipiens doubtless being so pronounced because of the already high levels of this fatty acid that are normally present in mosquitoes as in most Diptera (Fast 1970). This phenomenon is also very reminiscent of the increased levels of oleic-derived $\omega 9$-eicosatrienoic acid observed in mammals deprived of essential fatty acid which gives rise to the heightened triene/tetraene ratio.

It seems that although these various C18 polyunsaturate-deprived insects are biochemically anomalous, and in some cases develop at sub-optimal rates, the polyunsaturates that, when deficient, induce the synthesis of extra monoenoic acid, are not necessarily essential, over one generation at least, or over several in the mosquito case. It is perhaps helpful in this connection to distinguish between requirements that are essential (in the sense that, if not satisfied, development, reproduction, etc. would ultimately cease entirely), and those that merely optimize developmental rate (discussed in Dadd 1970). In the case of some Lepidoptera, linolenic acid is absolutely essential, since, without it, adults abort, whereas linoleic acid could be accounted an optimizing requirement only, since its absence apparently merely retards growth. It should be noted, however, that unless followed over sequential generations it is generally uncertain that a merely growth-optimizing requirement is no more than that.

In C. pipiens the ability to rear sequential generations with arachidonic acid as the only fatty acid in the diet suggests that although the absence of C18 or other polyunsaturated fatty acids causes the biochemical abnormality of unusually high palmitoleic acid, this does not necessarily entail their essentiality, though it suggests that in some way performances are likely to be suboptimal without them. Apart from mosquitoes, Diptera in general characteristically and normally synthesize high amounts of palmitoleic acid, and they have not been found to require any essential fatty acid in single generation studies, although non-essential fatty acids, particularly oleic, increased the growth rate of Agria affinis (House and Barlow 1960). Recalling that the essential fatty acid requirement in C. pipiens is extremely low, it seems possible that apart from a very small essential requirement for metabolic and/or microstructural functions, less specific structural fatty acid requirements, which in other insects such as Lepidoptera and Orthoptera would demand high amounts of dietary polyunsaturated fatty acids, may be accomodated by virtue of an ability to use biosynthesized monounsaturated fatty acids such as palmitoleic acid instead.

Were this the general case for Diptera, one can well imagine that a minute essential fatty acid requirement could easily pass undetected in the usual single generation experiment. It will be recalled that A. aegypti was long considered to require no fatty acid, yet it now appears to have the same need as other mosquitoes, though manifesting it clearly only later in development in terms of differences in adult longevity and fecundity. The housefly also gave no sign of a fatty acid requirement in standard nutritional studies (Brookes and Fraenkel 1958) and on fat-free diets apparently grows to the adult stage with tissues lacking detectable polyunsaturated fatty acids (Bridges 1970); yet with the recent detection of prostaglandin in its reproductive tissues (Wakayama et al. 1980) and its inability to biosynthesize linoleate (Blomquist et al. 1982), one must presume a need to acquire dietary, that is, essential, ω6 polyunsaturate.

Suppose we postulate, minimally, three physiological functions for polyunsaturated fatty acids, of which two, metabolic and microstructural, are absolutely essential, and the third is a less specific though quantitatively major structural function of a growth-optimizing nature that could be spared in whole or in part by monoenoic fatty acids; then it can be imagined that certain C20/C22 polyunsaturates, perhaps including both ω6 and ω3 members, while able to satisfy essential metabolic and microstructural functions, might be inadequate for the broader, optimizing structural function; this would be the case especially if the wider optimizing function included a need for C18 polyunsaturates and these could not be derived metabolically by retroconversion from the higher, absolutely essential polyunsaturates. The mosquito situation, and perhaps that of Diptera in general, would then be subsumed if the postulated broad, growth-optimizing structural function could be entirely spared by biosynthesized monoenoic fatty acids.

SUMMARY REMARKS

With so much still inconclusive, these comments necessarily have a tentative or outright speculative cast. Nevertheless, within their limitations even probabilistic assessments can help expose lacunae in current information and shape hypotheses for future and, one hopes, clarifying investigations.

On the issue of the generality of a dietary polyunsaturated fatty acid requirement among insects, the recently strengthened evidence for substantial C18 polyunsaturate biosynthesis in several species undermines, and some might say collapses, the idea of a universal dietary need. Three caveats should however be entered. If biosynthesis were localized in specific tissues, exogenous polyunsaturate might still be required. Next, were the physiological fatty acid requirement multiple, demanding, for instance, both ω6 and ω3 members, biosynthesis of one type only would still leave an exogenous dietary requirement for the others. Lastly, an element of uncertainty stems from the presence of symbiotes in most cases of possible polyunsaturate biosynthesis in the insect literature; however, in the recent and best documented study, species with an obligate symbiote-dependence occur among both linoleate synthesizers and non-synthesizers, which, on the face of it, suggests that microorganisms are not critical in this context.

Regardless of whether particular fatty acids are required in the diet or are dispensable because biosynthesized, the question of what polyunsaturates are ultimately needed for endogenous physiological purposes arises. In vertebrates the physiologically essential entities are C20 and C22 polyunsaturates, and if not ingested as such their derivation from shorter polyunsaturates available from food is well understood. The nutritional evidence from mosquitoes indicates that they too must physiologically require certain C20 or C22 polyunsaturates of a particular structure. For other insects, with or without a dietary C18 polyunsaturate requirement, the matter can at present be approached only obliquely. All those species most recently examined using gas chromatographic analysis contain C20 or C22 polyunsaturates in their tissues, suggesting a general physiological function for these fatty acids or their metabolites. The widespread occurrence of at least traces of long-chain polyunsaturates in insect tissues is especially compelling evidence for a functional role in the case of phytophages, which would be unlikely to acquire such fatty acids adventitiously from their plant food. Taken in conjunction with evidence for the sequestering of long-chain polyunsaturates in special tissues such as retinal membranes, and with the increasingly frequent detection of prostaglandins in insects of diverse taxa, which entails the presence of C20 fatty acid precursors, it currently seems most probable that insects generally, like vertebrates, will be found to have physiological needs for long-chain polyunsaturates. One such need will clearly revolve round the provision of prostaglandin-like metabolites. Though at present these are functionally implicated only in reproduction, the detection of prostaglandins in the brain and central nervous tissues of crickets and the evidence for an arachidonic acid/prostaglandin synthetase inhibitor effect in mosquitoes suggests a wider physiological involvement among insects. Apart from this, evidence for the sequestering of polyunsaturates in phospholipids and in particular membrane-rich tissues suggests that, as for vertebrates, a microstructural function in cellular membranes is likely, though there are currently no cytological and membrane-bound enzyme kinetic studies of fatty acid-deficient insect cells that might provide direct evidence of this.

Precisely which polyunsaturates may be necessary at both the physiological and dietary levels seems likely to differ substantially among insect species as it does among vertebrates. The quite different structure/function relations for fatty acids that can satisfy the respective requirements of Culex pipiens and Galleria mellonella heavily underscores this. This apart, among both vertebrate and insects linolenate (or ω3) specialists can be distinguished. Warm-blooded vertebrates require primarily ω6 fatty acids, and while the available evidence does not distinguish a category of primarily ω6-requiring insects in the positive way that many Lepidoptera are clearly ω3-dependent, several insect species can apparently satisfy their essential fatty acid requirement completely with linoleic acid alone.

It is altogether possible that fatty acids of the ω6 and ω3 families may both be essential for many insects, the evidence for this being particularly persuasive for many Lepidoptera if account is taken of both the pupal/ adult emergence requirement and the retarded larval growth in the absence of both C18 fatty acids. Two groups of findings provide the most telling evidence for this possibility: the prostaglandins so far found in insects are predominantly of the 2 series derived from arachidonic acid (ω6), including those of the lepidop- teran silk moth for which linolenic acid (ω3) is the main required dietary fatty acid; and retinal phospho- lipid fatty acids of insects of 3 orders, including a plant-eating lepidopteran, contain substantial propor- tions of both arachidonic and eicosapentaenoic acids, respectively ω6 and ω3. It is interesting that a double requirement for both ω6 and ω3 polyunsaturates is considered a possibility among vertebrates also.

REFERENCES

Alfin-Slater, R.B. and Aftergood, L. (1971) Physiologi- cal function of essential fatty acids. Progr. Bio- chem. Pharmacol. 6: 214-244.
Bade, M.L. (1964) Biosynthesis of fatty acids in the roach Eurycotis floridana. J. Insect Physiol. 10: 333-341.
Blomquist, G.J., Dwyer, L.A., Chu, A.J., Ryan, R.O. and de Renobales, M. (1982) Biosynthesis of linoleic acid in a termite, cockroach and cricket. Insect Biochem. 12: 349-353.
Brookes, V.J. and Fraenkel, G. (1958) The nutrition of the larva of the housefly, Musca domestica L. Physi- ol. Zool. 31: 208-223.
Bridges, R.G. (1971) Incorporation of fatty acids into the lipids of the housefly, Musca domestica. J. In- sect Physiol. 17: 881-895.
Chippendale, G.M. (1971) Lipid requirements of the angoumois grain moth, Sitotroga cerealella. J. In- sect Physiol. 17: 2169-2177.
Chippendale, G.M., Beck, S.D. and Strong, F.M. (1964) Methyl linolenate as an essential nutrient for the cabbage looper, Trichoplusia ni (Hübner). Nature 204: 710-711.
Cohen, E. (1974) Fatty acid synthesis by the hide beetle Dermestes maculatus (Dermestidae: Coleoptera). Ent. Exp. Appl. 17: 433-438.
Dadd, R.H. (1961) The nutritional requirements of locusts--V. Observations on essential fatty acids, chlorophyll, nutritional salt mixtures, and protein or amino acid components of synthetic diets. J. In-

142

sect Physiol. 6: 126-145.
Dadd, R.H. (1964) A study of carbohydrate and lipid nutrition in the wax moth, Galleria mellonella (L.), using partially synthetic diets. J. Insect Physiol. 10: 161-178.
Dadd, R.H. (1970) Arthropod Nutrition, in Chemical Zoology, vol. V. Editors: M. Florkin and B.T. Scheer (Academic Press, New York).
Dadd, R.H. (1973) Insect nutrition: current development and metabolic implications. Annu. Rev. Ent. 18: 331-420.
Dadd, R.H. (1977) Qualitative requirements and utilization of nutrients: Insects, in Handbook series in nutriton and food, vol. 1, pp. 305-346. Editor: M. Rechcigl (CRC Press, Cleveland).
Dadd, R.H. (1980) Essential fatty acids for the mosquito Culex pipiens. J. Nutr. 110: 1152-1160.
Dadd, R.H. (1981) Essential fatty acids for mosquitoes, other insects and vertebrates, in Current topics in insect endocrinology and nutrition, pp. 184-214. Editors: G. Bhaskaran, S. Friedman, and J.G. Rodriguez (Plenum, New York).
Dadd, R.H. and Kleinjan, J.E. (1976) Chemically defined dietary media for larvae of the mosquito Culex pipiens (Diptera: Culicidae): effects of colloid texturizers. J. med. Ent. 13: 285-291.
Dadd, R.H. and Kleinjan, J.E. (1978) An essential nutrient for the mosquito Culex pipiens associated with certain animal-derived phospholipids. Ann. Ent. Soc. Amer. 71: 794-800.
Dadd, R.H., and Kleinjan, J.E. (1979a) Essential fatty acid for the mosquito Culex pipiens: Arachidonic acid. J. Insect Physiol. 25: 495-502.
Dadd, R.H. and Kleinjan, J.E. (1979b) Vitamin E, ascorbyl palmitate or propyl gallate protect arachidonic acid in synthetic diets for mosquitoes. Ent. Exp. Appl. 26: 222-226.
Dadd, R.H., Kleinjan, J.E., and Sneller, V.-P. (1977) Development of several species of mosquito larvae in fully defined dietary media: preliminary evaluation. Mosquito News 37: 699-703.
Dadd, R.H., Friend, W.G., and Kleinjan, J.E. (1980) Arachidonic acid requirement for two species of Culiseta reared on a synthetic diet. Canad. J. Zool. 58: 1845-1850.
Davis, G.R.F. and Sosulski, F.W. (1973) Improvement of basic diet for use in determining the nutritional value of proteins with larvae of Tenebrio molitor L. Arch. Int. Physiol. Biochim. 81: 495-500.
Destephano, D.B. and Brady, V.E. (1977) Prostaglandin and prostaglandin synthetase in the cricket, Acheta domesticus. J. Insect Physiol. 23: 905-911.

Downer, R.G.H. (1978) Functional role of lipids in insects, in Biochemistry of Insects. Editor: M. Rockstein (Academic Press, New York).

Dwyer, L.A. and Blomquist, G.J. (1981) Biosynthesis of linoleic acid in the cockroach, P. americana. Proceedings of the 25th Jubilee Congress on essential fatty acids and prostaglandins. Progr. Lipid Res. 20: 215-218.

Earle, N.W., Slatten, B. and Burks, M.L. (1967) Essential fatty acids in the diet of the boll weevil, Anthonomus grandis Boheman. J. Insect Physiol. 13: 187-200.

Fast, P.G. (1964) Insect lipids: a review. Memoirs Ent. Soc. Canad. No. 37.

Fast, P.G. (1967) An analysis of the lipids of Gryllus bimaculatus (De Geer) (Insecta, Orthoptera). Canad. J. Biochem. 45: 503-505.

Fast, P.G. (1970) Insect lipids. Progr. Chem. Fats Lipids 11: 181-242.

Fawzi, M., Osman, M.F.N. and Schmidt, G.H. (1961) Analyse der Körperfetter von imaginalen Wanderheuschrecken der art Locusta migratoria migratorioides L. Biochem. Z. 334: 441-450.

Fraenkel, G. and Blewett, M. (1946) Linoleic acid, vitamin E and other fat-soluble substances in the nutrition of certain insects (Ephestia kuehniella, E. elutella, E. cautella and Plodia interpunctella (Lep.)). J. Exp. Biol. 22: 172-190.

Fraenkel, G. and Blewett, M. (1947) Linoleic acid and arachidonic acid in the metabolism of the insects Ephestia kuehniella and Tenebrio molitor. Biochem. J. 41: 475-478.

Friend, W.G. (1968) The nutritional requirements of Diptera, in: Radiation, radioisotopes and rearing methods in the control of insect pests. Proc. IAEA, Vienna, pp. 41-57.

Galli, C., Spagnuolo, G., Agradi, E. and Paoletti, R. (1976) Comparataive effects of olive oil and other edible fats on brain structural lipids during development. Lipids 1: 237-243.

Gilbert, L.I. (1967) Lipid metabolism and function in insects. Adv. Insect Physiol. 4: 69-211.

Gilbert, L.I. and O'Connor, J.D. (1970) Lipid metabolism and transport in arthropods, in: Chemical Zoology, vol. 5. Editors: M. Florkin and B.T. Scheer (Academic Press, New York).

Golberg, L. and DeMeillon, B. (1948) The nutrition of the larvae of Aedes aegypti Linnaeus. 3. Lipid requirements. Biochem. J. 43: 372-379.

Gordon, H.T. (1959) Minimal nutritional requirements of the German roach, Blattella germanica L. Ann. N.Y. Acad. Sci. 77: 290-351.

Grau, P.A. and Terriere, L.C. (1971) Fatty acid profiles of the cabbage looper, Trichoplusia ni, and
the effect of diet and rearing conditions. J. Insect
Physiol. 17: 1637-1649.

Guarnieri, M. and Johnson, R.M. (1970) The essential
fatty acids. Adv. Lipid Res. 8: 115-174.

Holman, R.T. (1967) Essential fatty acid deficiency--a
long scaly tail. Progr. Chem. Fats Lipids 6: 279-
348.

Holman, R.T. (1977) The deficiency of essential fatty
acids, in Polyunsaturated fatty acids, pp. 163-182.
Editors: W.-H. Kunau and R.T. Holman (American Oil
Chemistry Society, Champaign).

House, H.L. and Barlow, J.S. (1960) Effects of oleic
and other fatty acids on growth rate of Agria
affinis (Fall.) (Diptera: Sarcophagidae). J. Nutr.
72: 409-414.

Ito, T. and Nakasone, S. (1966) Nutrition of the silkworm, Bombyx mori. XIII. Nutritive effects of
fatty acids. Bull. Sericult. Exp. Station 20:
375-391.

Jenkin, H.M., McMeans, E., Anderson, L.E. and Yang, T.K.
(1975) Comparison of phospholipid composition of
Aedes aegypti and Aedes albopictus cells from
logarithmic and stationary phases of growth. Lipids
10: 686-694.

Jenkin, H.M., McMeans, E., Anderson, L.E. and Yang, T.K.
(1976) Phospholipid composition of Culex quinquefasciatus and Culex tritaeniorrhyncus cells in logarithmic and stationary growth phases. Lipids 11: 697-704.

Keith, A.D. (1967) Fatty acid metabolism in Drosophila
melanogaster: interaction between dietary fatty
acids and de novo synthesis. Comp. Biochem.
Physiol. 21: 587-600.

Kinsella, J.E. (1966) Metabolic patterns of the fatty
acids of Periplaneta americana (L.) during its
embryonic development. Canad. J. Biochem. 44:
247-258.

Kircher, H.W. and Gray, M.A. (1978) Cholestanol-choles-
terol utilization by axenic Drosophila melanogaster.
J. Insect Physiol. 24: 555-559.

Kleinjan, J.E. and Dadd, R.H. (1977) Vitamin requirements of the larval mosquito, Culex pipiens. Ann.
Ent. Soc. Amer. 70: 541-543.

Lands, W.E.M., Martin, E.H. and Crawford, C.G. (1977)
Functions of polyunsaturated fatty acids:
biosynthesis of prostaglandins, in Polyunsaturated
fatty acids, pp. 193-228. Editors: W.-H. Kunau and
R.T. Holman (American Oil Chemists Society,
Champaign).

Loher, W. (1979) The influence of prostaglandin E_2 on
oviposition in Teleogryllus commodus. Ent. Exp.

Appl. 25: 107-108.

Loher, W., Ganjian, I., Kubo, I., Stanley-Samuelson, D. and Tobe, S.S. (in press). Prostaglandins: their role in egg-laying of the cricket, Teleogryllus commodus. Proc. Nat. Acad. Sci. 78: 7835-7838.

Louloudes, S.J., Kaplanis, J.N., Robbins, W.E. and Monroe, R.E. (1961) Lipogenesis from C^{14}-acetate by the American cockroach. Ann. Ent. Soc. Amer. 54: 99-103.

Mead, J.F. (1970) The metabolism of the polyunsaturated fatty acids. Progr. Chem. Fats Lipids 9: 161-192.

Meikle, J.E.S. and McFarlane, J.E. (1965) The role of lipid in the nutrition of the house cricket, Acheta domesticus L. (Orthoptera: Gryllidae). Canad. J. Zool. 43: 87-98.

Morère, M. J.-L. (1971) Remplacement de l'huile de germe de mäis par des acides gras dans l'alimentation de type méridique (= semi-synthetique) de Plodia interpunctella (Hbn.) (Lepidoptère-Pyralidae). C.R. Acad. Sc. Paris 272: 133-136.

Nakasone, S. and Ito, T. (1967) Fatty acid composition of the silkworm, Bombyx mori L. J. Insect Physiol. 13: 1237-1246.

Nayar, J.K. (1964) The nutritional requirements of grasshoppers. I. Rearing of the grasshopper, Melanoplus bivittatus (Say), on a completely defined synthetic diet and some effects of different concentrations of B-vitamin mixture, linoleic acid, and ß-carotene. Canad. J. Zool. 42: 11-22.

Rivers, J.P.W., Sinclair, A.J. and Crawford, M.A. (1975) Inability of the cat to desaturate essential fatty acids. Nature 258: 171-173.

Rock, G.D. (1967) Aseptic rearing of the codling moth on synthetic diets: ascorbic acid and fatty acid requirements. J. Econ. Ent. 60: 1002-1005.

Rock, G.D., Patton, R.L. and Glass, E.H. (1965) Studies on the fatty acid requirements of Argyrotaenia velutinana (Walker). J. Insect Physiol. 11: 91-101.

Sang, J.H. (1978) The nutritional requirements of Drosophila, in: "The genetics and Bilogy of Drosophila" Volume 2, Editors: M. Ashburner and T.R.F. Wright (Academic Press, New York).

Schaefer, C.H. and Washino, R.K. (1969) Changes in the composition of lipids and fatty acids in adult Culex tarsalis and Anopheles freeborni during the overwintering period. J. Insect Physiol. 15: 395-402.

Schaefer, C.H. and Washino, R.K. (1970) Synthesis of energy for overwintering in natural populations of the mosquito Culex tarsalis. Comp. Biochem. Physiol. 35: 503-506.

Schmidt, G.H. and Osmun, M.F.H. (1962) Analyse des

Raupenöls von Mondvögel Phalera bucephala L. (Lepidoptera: Notodontidae). J. Insect Physiol. 8: 233-240.

Setty, B.N.Y. and Ramaiah, T.R. (1980) Effect of prostaglandins and inhibitors of prostaglandin biosynthesis on oviposition in the silkmoth Bombyx mori. Indian J. Exp. Biol. 18: 539-541.

Sinclair, H.M. (1964) Essential fatty acids, in Lipid Pharmacology, pp. 237-273 (Academic Press, London).

Singh, K.R.P. and Brown, A.W.A. (1957) Nutritional requirements of Aedes aegypti L. J. Insect Physiol 1: 199-220.

Sivapalan, P. and Gnanapragasam, N.-C. (1979) The influence of linoleic and linolenic acid on adult moth emergence of Homona coffearia from meridic diets in vitro. J. Insect Physiol. 25: 393-398.

Sneller, V.-P. and Dadd, R.H. 1981. Lecithin in synthetic larval diet for Aedes aegypti improves larval and adult performance. Ent. Exp. Appl. 29: 9-18.

Stanley-Samuelson, D. (1980) Long-chain polyunsaturated fatty acids in whole-animal and specific-tissue extracts of insects. Abstract #577, Am. Zool. 20: 832.

Stanley-Samuelson, D. and Dadd, R.H. (1981) Arachidonic and other tissue fatty acids of Culex pipiens reared with various concentrations of dietary arachidonic acid. J. Insect Physiol. 27: 571-578.

Stephen, W.F. and Gilbert, L.I. (1969) Fatty acid biosynthesis in the silkmoth, Hyalophora cecropia. J. Insect Physiol. 15: 1833-1854.

Strong, F.E. (1963) Fatty acids: in vivo synthesis by the green peach aphid, Myzus persicae (Sulzer). Science 140: 983-984.

Tamaki, Y. (1961) Studies on nutrition and metabolism of the smaller tea tortrix, Adoxophyes orana (Fisher von Rösterstamm). II. An essential factor for adult emergence. Jap. J. Appl. Ent. Zool. 5: 58-63.

Terriere, T.C. and Grau, P.A. (1972) Dietary requirements and tissue levels of fatty acids in three Noctuidae. J. Insect Physiol. 18; 633-647.

Thompson, S.N. (1981) The nutrition of parasitic Hymenoptera. Proc. IXth Int. Congr. Plant Protection 1: 93-96.

Tinoco, J., Babcock, R. Hincenbergs, I., Medwadowski, B., Miljanich, P. and Williams, M.A. (1979) Linolenic acid deficiency. Lipids 14: 166-173.

Turunen, S. (1973) Utilization of fatty acids by Pieris brassicae reared on artificial and natural diets. J. Insect Physiol. 19: 1999-2009.

Turunen, S. (1974) Polyunsaturated fatty acids in the nutrition of Pieris brassicae (Lepidoptera). Ann. Zool. Fennici 11: 300-303.

Turunen, S. (1976) Vitamin E: effect on lipid synthesis

and accumulation of linolenate in _Pieris brassicae._ Ann. Zool. Fennici 13: 148-152.

Vanderzant, E.S., Kerur, D. and Reiser, R. (1967) The role of dietary fatty acids in the development of the pink bollworm. J. Econ. Ent. 50: 606-608.

Wakayama, E.J., Dillwith, J.W. and Blomquist, G.J. (1980) In vitro biosynthesis of prostaglandins in the reproductive tissues of the male housefly, _Musca domestica_ L. Abstract #1010, Am. Zool. 20: 904.

Walling, M.V., White, D.C. and Rodriguez, J.C. (1968) Characterization, distribution, catabolism, and synthesis of the fatty acids of the two-spotted spider mite, _Tetranychus urticae._ J. Insect Physiol. 14: 1455-1458.

Worthington, R.E., Brady, V.E., Thean, J.E. and Wilson, D.M. (1981) Arachidonic acid: occurrence in the reproductive tract of the male house cricket (_Acheta domesticus_) and field cricket (_Gryllus_ sp.). Lipids 16: 79-81.

Yazgan, S. (1972) A chemically defined synthetic diet and larval nutritional requirements of the endoparasitoid _Itoplectis conquisitor_ (Hymenoptera). J. Insect Physiol. 18: 2123-2141.

Yone, Y. and Fujii, M. (1975) Studies on nutrition of red sea bream--XII. Effect of ω3 fatty acid supplement in a corn oil diet on fatty acid composition of fish. Bull. Jap. Soc. Sci. Fish 41: 79-86.

Zinkler, D. (1975) Zum Lipidmuster der Photorezeptoren von Insekten. Verh. Dtsch. Zool. Ges. 67: 28-32.

8
Lipid Factors in Insect Growth and Reproduction

*J. E. McFarlane**

INTRODUCTION

This article will deal mainly with the work in our laboratory on the physiology of the fat-soluble vitamins E and K_1 in the house cricket, Acheta domesticus.

Dietary vitamin E was first shown to improve growth of an insect in 1946 by Fraenkel and Blewett, working with the lepidopterans, Anagasta kuehniella and A. elutella. Fraenkel and Blewett also found that vitamin E acted as an anti-oxidant for the polyunsaturated fatty acid linoleic acid, which itself improved growth and was required for the emergence of normal adults. The antioxidant rôle of vitamin E was fulfilled by other antioxidants, i.e. ascorbic acid and propyl gallate, but only vitamin E improved growth when presented alone. Fraenkel and Blewett considered that vitamin E improved growth probably by acting as a physiological antioxidant for body lipid.

For the next 16 years extensive nutritional work with insects showed no requirement for vitamin E. It was in fact generally believed during this period that fat-soluble vitamins were not required by insects, and that this difference in requirements between insects and vertebrates indicated basic differences in the physiology of the two groups. However, since the early sixties, the only fat-soluble vitamin found to be without a beneficial effect on some insect species or other is vitamin D. Vitamin A functions as a component of rhodopsin in insects as well as in mammals (Needham 1978).

Vitamin E has been shown to be required for reproduction in three species of insects from three orders. Chumakova (1962) found that vitamin E was essential for reproduction in the beetle Cryptolaemus montrouzieri. Meikle and McFarlane (1965) demonstrated conclusively

* Department of Entomology, Macdonald Campus of McGill University, Ste. Anne de Bellevue, P.Q., Canada. Study supported by a grant from the National Sciences and Engineering Research Council of Canada.

that vitamin E was required for spermatogenesis in the orthopteran Acheta domesticus, and House (1966) showed that vitamin E was required for ovarian development in the dipteran Agria housei. The only other invertebrates which have been shown to require vitamin E are the rotifers (Gilbert, 1974); in this group, the growth and reproduction of certain species are affected.

In addition to these reproductive effects, vitamin E improves growth in Acheta and Agria. It appears that whereas the reproductive effect in Acheta is due to the ring structure of α-tocopherol, and hence is probably due to the anti-oxidant property of the molecule, the growth effect seems to be due to the phytyl side chain (McFarlane, 1976). This distinction between ring and side chain effects in Acheta may help to clarify the rôle of vitamin E in other eukaryotes.

REPRODUCTION

To consider firstly the reproductive effect of vitamin E, Acheta reared on diets lacking E do not produce viable eggs, or produce them in very low quantity (Meikle and McFarlane, 1965). The threshold for 'normal' viable egg production is about 17.5 µg/g. Above this level there is little or no improvement in viable egg production, whereas below it fertility falls off sharply (McFarlane, 1972).

The reproductive effect of vitamin E deficiency in Acheta is to prevent spermiogenesis (Fig. 1), that is, the transformation of spermatids to spermatozoa (Meikle and McFarlane, 1965). In this respect, the effect is similar to the effect on the rat, although in the rat there are other histopathological effects, such as degeneration of the epithelium of the seminiferous tubule (Mason, 1954). The usual explanation for these and indeed all other effects of vitamin E deficiency is that peroxidation of polyunsaturated fatty acids (PUFA) occurs with consequent structural and functional damage to subcellular membranes (McCay et al., 1972).

One of the unusual aspects of house cricket nutrition, when compared with that of many other insects, is that the house cricket grows reasonably well on a diet lacking fatty acids, and yet can form linoleic acid, a PUFA that is essential for many insects (Meikle and McFarlane, 1965). Recently, Blomquist et al. (1981) have demonstrated the in vitro synthesis of linoleic acid by house cricket tissues. However, crickets without fatty acids are sterile unless α-tocopherol is present in the diet. Evidently the antioxidant rôle is crucial for the testis, but not for other tissues.

Probably as a result of the effect on spermatogenesis, the size of the testis is reduced in vitamin E-deficient insects, but significantly only in the last larval instar (Meikle and McFarlane, 1965). Withholding vitamin E from the larval stage or from the imago has shown that spermatogenesis occurs largely, if not entirely, in the larval stage. Further work has narrowed spermatogenesis to the last larval stadium. This is the stadium in which juvenile hormone is absent, although the molting hormone, ecdysone, is still required to produce the final molt (Riddiford and Truman, 1978). There may be a relationship, possible steric, between juvenile hormone and vitamin E which affects testis development. It would be useful to study these possible interactions with testes maintained <u>in</u> <u>vitro</u>.

Vitamin E, α-tocopherol quinone and vitamin K_1 are structural analogues, having the phytyl side-chain in common (Fig. 2). Modification of the ring abolishes the reproductive effect, neither α-tocopherol quinone (McFarlane, 1972b) nor vitamin K_1 being effective (McFarlane, 1976b). In view of the fact that selenium can replace vitamin E for some of the functions of the vitamin (Schwarz, 1965), the possibility of substituting selenium for vitamin E was investigated (McFarlane, 1972b). Selenium does not replace vitamin E for its reproductive function.

REPRODUCTIVE PHYSIOLOGY OF THE MALE CRICKET

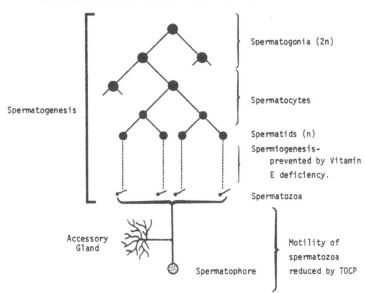

FIGURE 1.

152

Vitamin E-deficient males, then, do not form spermatozoa, so there is no transfer of sperm to the spermatheca of the female. Uninseminated females show a longer
preoviposition period and lay fewer eggs than inseminated
females (Meikle and McFarlane, 1965). The reason for
this is not clear, but possibly is due to an effect on
prostaglandins. This will be discussed later.

The basic diet used in our experiments has the following composition: casein 40g, cellulose fiber 30g, d(+)
dextrose 20g, salt mixture 4g, cholesterol 1g, and a mixture of B-vitamins (McFarlane, 1978). The salt mixture
used in our earlier nutritional experiments was U.S.P.
XIV at a level of 2g/93g diet. Adult males deficient in
cuticular melanin were found among insects reared on this
basic diet with vitamin E (McFarlane, 1972a). These
adult males were designated albinos, and although sclero-
tization occurred normally and there was a slight amber
coloration of the cuticle, the pigmentation was markedly
less than normal. This pointed to an involvement of
vitamin E with Cu metabolism, as the oxidative enzymes

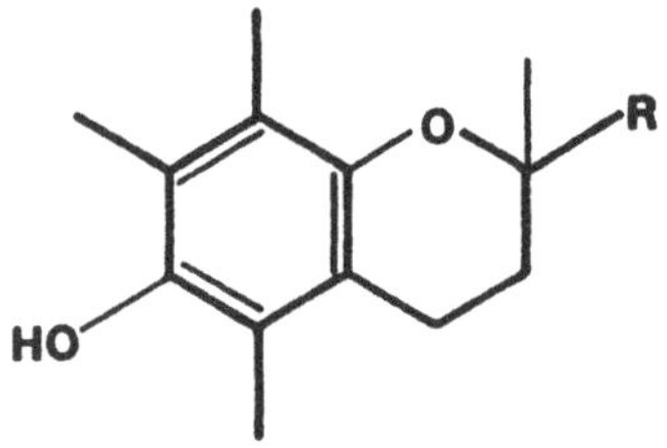

Vitamin E
(α-Tocopherol)

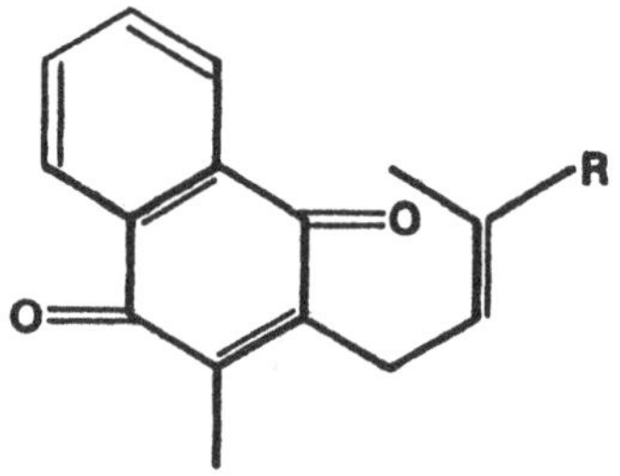

Vitamin K₁

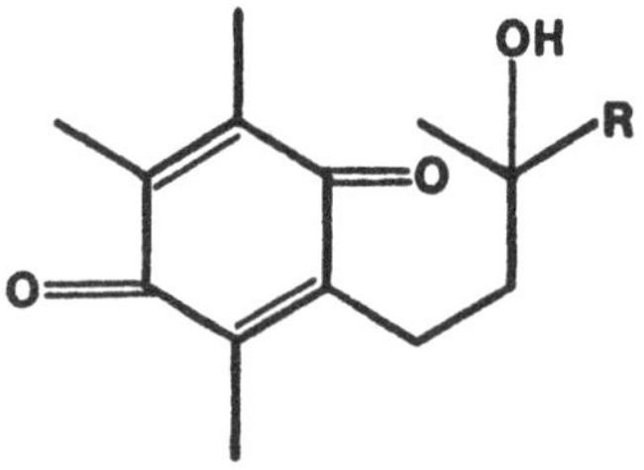

α-Tocopherol quinone

= R

FIGURE 2.

leading to melanin formation contain Cu. Increasing the amount of Cu in the diet abolished the albino effect (McFarlane, 1974). The U.S.P. salt mixture would have provided added Cu at a level of 0.66 µg/g diet. It therefore appeared that at low levels of Cu in the diet, Cu might be incorporated preferentially into the testis rather than being used to form cuticular tyrosinase. Examination of the Cu level in the testis (McFarlane, 1974) showed it to be much higher than in the food (16X) or in the whole body (8X). Transferring albino males reared on artificial diet to chow diets containing a higher level of Cu did not darken the cuticle further (McFarlane, 1972a). Moreover, isolated albino cuticle incubated with tyrosine also did not darken further. Evidently tyrosine is not formed in the cuticle during the last larval stadium of what will become an albino adult male. The pigmentation of immature stadia and of adult females is unaffected (McFarlane, 1978).

Cu present as an impurity in the casein gives a level of 2 µg Cu/g basic diet. We have not tried to reduce the Cu level further by extraction of the casein. However, reproduction in the presence of vitamin E is variable on this diet, whereas it is consistently normal when 6 µg/g of Cu or more are added (McFarlane, 1976). Thus it appears that Cu is required for formation of sperm in adequate quantities, and that the reproductive function takes precedence over the cosmetic function of pigmentation. The precise relationship, if any, between vitamin E and Cu in reproduction is not yet understood.

In this connection, it is interesting that Distler and McFarlane (1981) have reported that where dietary Cu is high, dietary linoleic acid prevents spermiogenesis, an effect identical to that found in vitamin E deficiency. This is thought to be due to catalytic peroxidation in the gut of the linoleic acid, which would reduce the vitamin E content to a point where the males are effectively sterilized.

GROWTH

Turning now to growth, an effect of vitamin E on the growth of the male can only be demonstrated at adequate levels of dietary Cu (McFarlane, 1974). This does not necessarily mean that the relationship between E and Cu is the same in growth as in reproduction. As shown in the preceding section, the ring structure of E is crucial for reproduction. Vitamin K_1, which has the same phytyl side chain as E but is not an anti-oxidant, also stimulates male growth, and the effects of E and K_1 have been shown to be additive in the female (McFarlane, 1978). As vitamin K_1 produces its effect in the male at a lower level than E (1.87 µg/g diet as

against 17.2 µg/g diet), it may have a sparing action on
E. Various sythetic vitamins K and the K analogue coumarin (Fig. 3), which do not have the phytyl side chain,
have no effect on growth (McFarlane, 1976b). That vitamin K may be a growth factor for birds and mammals is a
possibility that has never been considered, because of
the drastic effect of vitamin K deficiency resulting in
hemorrhagic disease (Olson and Suttie, 1977). Vitamin K
has not previously been shown to be a growth factor of
eukaryotic organisms.

CHEMOSTERILIZATION

Consideration has been given to the possibility of
producing sterile males with a view to insect control.
While it is possible to produce sterile males by rearing
crickets on an artificial diet without vitamin E, the
method is expensive, and the resulting adult is considerably smaller than males reared on a rabbit chow diet, the
diet used for the stock culture of the insect. We therefore tried adding a series of vitamin E antagonists to
the chow diet with a view to obtaining sterile males of
a greater weight and perhaps competitiveness. Of five
antagonists used, only one, tri-o-cresyl phosphate (TOCP)
(Fig. 4) gave sterilization at levels which were not
otherwise very toxic to the insect (Prévost and McFarlane, 1980a). A slight but significant effect on growth
was observed, which was not unexpected as vitamin E affects growth.

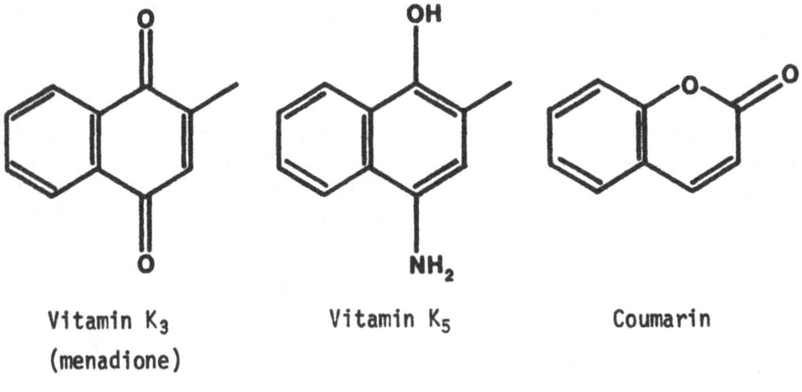

Vitamin K₃
(menadione)

Vitamin K₅

Coumarin

FIGURE 3.

It was rather surprising to find that TOCP-treated males produced normal amounts of spermatozoa (Prévost and McFarlane, 1980b). It is possible that TOCP, unlike E, was not able to penetrate a blood-testis barrier. Such a barrier has been shown to exist in Locusta migratoria in the region of the differentiating germ cells (Szollosi and Marcailloux, 1977). Furthermore, the sperm were transferred in normal amounts to the spermatheca of the female. The only effect observed was on sperm motility: spermatophoral sperm and spermathecal sperm were much less motile than normal sperm. Adding vitamin E prevented the effects of TOCP. These results suggest that vitamin E has a second rôle in the reproduction of the house cricket male, which is to maintain sperm motility.

We can only speculate at the present time as to a mechanism. Peroxidation of fowl sperm lipid has been shown to inhibit motility (Fujihara and Howarth, 1978). There may be an effect through prostaglandins as well. A substance resembling prostaglandin E_2 has been shown to be present in the testes and male reproductive tract of Acheta domesticus, and in the reproductive tissues of mated females, by Destephano and Brady (1977). Although these authors consider E_2 acts in the female to increase egg production, E_2 may have an effect on sperm motility as well. Prostaglandins in the rabbit may function to preserve sperm motility (Spilman et al., 1973). Vitamin E may ensure the stability of prostaglandins and their precursor, arachidonic acid. In this regard it may be significant that the preoviposition period was lengthened and fecundity of females fed TOCP was reduced (Prévost and McFarlane, 1980b). As noted previously, vitamin E-deficient females mated to vitamin E-deficient males also showed similar effects on the preoviposition

Tri-o-cresyl phosphate

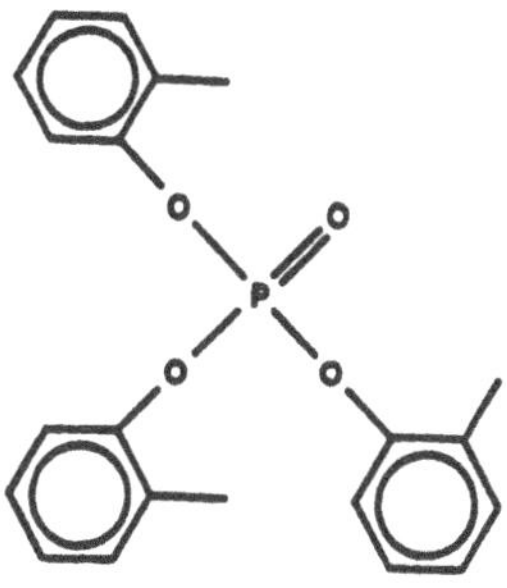

FIGURE 4.

156

period and on fecundity, even though no sperm were transferred (Meikle and McFarlane, 1965). In both instances,
the effects may be due to the absence or a low level of
prostaglandins.

REFERENCES

Blomquist, G.J., L.A. Dwyer, A.J. Chu, R.O. Ryan and M.
    de Renobales (1982) Biosynthesis of linoleic acid
    in a termite, cockroach and cricket. Insect Biochem.
    12: 349-353.
Chumakova, B.M. (1962) Significance of individual food
    components for the vital activity of mature predatory
    and parasitic insects. Vop. Ekol. Kievsk. 8: 133-
    134. (Biol. Abstr. 45: 44, 502, 1964).
Destephano, D.B. and U.E. Brady (1977) Prostaglandin
    and prostaglandin synthetase in the cricket, Acheta
    domesticus. J. Insect Physiol. 23: 905-911.
Distler, M.H.W. and J.E. McFarlane (1981) Inhibition of
    spermiogenesis by dietary linoleic acid in the house
    cricket, Acheta domesticus (L.). Comp. Biochem.
    Physiol. (in press).
Fraenkel, G. and M. Blewett (1946) Linoleic acid, vita-
    min E and other fat soluble substances in the nutri-
    tion of certain insects, Ephestia kuehniella, E. lu-
    tella, E. cautella, and Plodia interpunctella. J.
    Exper. Biol. 22: 172-190.
Fujihara, N. and B. Howarth, Jr. (1978) Lipid peroxida
    tion in fowl spermatozoa. Poultry Sci. 57: 1766-
    1768.
Gilbert, J.J. (1974) Effect of tocopherol on the growth
    and development of rotifers. Am. J. Clin. Nutr. 27:
    1005-1016.
House, H.L. (1966) Effects of vitamins E and A on
    growth and development, and the necessity of vitamin
    E for reproduction in the parasitoid Agria affinis
    (Fallén) (Diptera, Sarcophagidae). J. Insect
    Physiol. 12: 409-417.
Mason, K.E. (1954) Physiological action of vitamin E
    and its homologues. In The Vitamins, vol. 2, eds.
    W.H. Sebell, Jr. and H.S. Harris, pp. 514-562.
    Academic Press, New York.
McCay, P.B., Pfeifer, P.M. and W.H. Stripe (1972) Vita-
    min E protection of membrane lipids during electron
    transport functions. Ann. N.Y. Acad. Sci. 203: 72-
    73.
McFarlane, J.E. (1972a) Studies on vitamin E in the
    house cricket, Acheta domesticus (Orthoptera: Gryl-
    lidae) I. Nutritional albinism. Can. Ent. 104: 511-
    514.
McFarlane, J.E. (1972b) Vitamin E, tocopherol quinone
    and selenium in the diet of the house cricket, Acheta

domesticus. Israel J. Ent. 7: 7-14.

McFarlane, J.E. (1974) The functions of copper in the house cricket and the relation of copper to vitamin E. Can. Ent. 106: 441-446.

McFarlane, J.E. (1976a) Influence of dietary copper and zinc on growth and reproduction of the house cricket (Orthoptera: Gryllidae). Can. Ent. 108: 387-390.

McFarlane, J.E. (1976b) Vitamin K: a growth factor for the house cricket (Orthoptera: Gryllidae). Can. Ent. 108: 391-394.

McFarlane, J.E. (1978) Vitamins E and K in relation to growth of the house cricket (Orthoptera: Gryllidae). Can. Ent. 110: 329-330.

Meikle, J.E.S. and J.E. McFarlane (1965) The role of lipid in the nutrition of the house cricket, Acheta domesticus (L.) (Orthoptera: Gryllidae). Can. J. Zool. 43: 87-98.

Needham, A.E. (1978) Insect biochromes: their chemistry and role. In Biochemistry of Insects, ed. M. Rockstein, pp. 233-305. Academic Press, New York.

Olson, R.E. and J.W. Suttie (1977) Vitamin K and α-carboxyglutamate synthesis. Vit. & Hormon. 35: 59-108.

Prévost, Y.H. and J.E. McFarlane (1980a) Evaluation of tri-o-cresyl phosphate as an insect chemosterilant. Ann. Ent. Soc. Que. 25: 27-35.

Prévost, Y.H. and J.E. McFarlane (1980b) The mode of action of tri-o-cresyl phosphate, a house cricket chemosterilant. Ent. exp. applic. 27: 50-58.

Riddiford, L.M. and J.W. Truman (1978) Biochemistry of insect hormones and insect growth regulators. In Biochemistry of Insects, ed. M. Rockstein, pp. 308-357. Academic Press, New York.

Schwarz, K. (1965) Role of vitamin E, selenium and related factors in experimental nutritional liver disease. Fed. Proc. 24: 58-67.

Spilman, C. H., Fin, A. E. and J. F. Norland (1973) Effects of prostaglandins on sperm transport and fertilization in the rabbit. Prostaglandins 4: 57-63.

Szollosi, A. and C. Marcaillou (1977) Electron microscope study of the blood/testis barrier in an insect, Locusta migratoria. J. Ultrastruct. Res. 59: 158-172.

9
Insect Phospholipids

R. G. Bridges*

STRUCTURE OF PHOSPHOLIPIDS

Phospholipids are loosely defined as compounds which contain phosphorus and which are soluble in non-polar solvents. Compounds, extractable from natural products, which fall into this definition can be divided into two groups; one based on compounds containing glycerol--the phosphoglycerides, and the other on compounds containing sphingosine--the phosphosphingo-lipids.

Most of the compounds in the first group have the general formula:

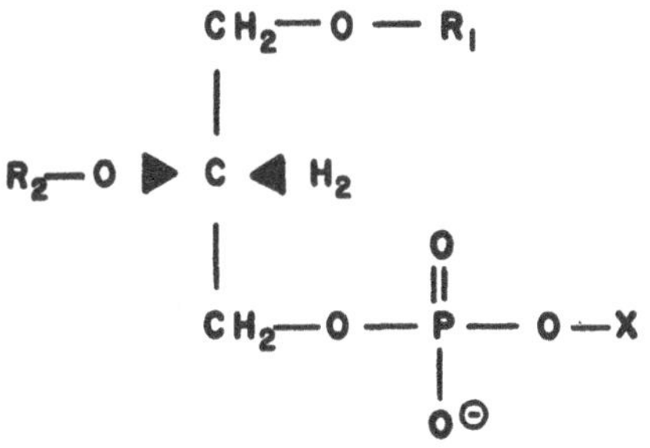

$$CH_2 - O - R_1$$
$$R_2 - O \blacktriangleright C \blacktriangleleft H_2$$
$$CH_2 - O - \overset{\displaystyle O}{\underset{\displaystyle O^{\ominus}}{\overset{\displaystyle \|}{P}}} - O - X$$

R_1 and R_2 are long chain acyl-, alkylor alk-1-enyl-groups usually of between 14 to 20 carbons atoms. X can either be an ethanolamine, choline, 1-serine or myo-inositol residue, or simply -H if phosphatidic acid. If the

* A.R.C. Unit of Invertebrate Chemistry and Physiology, Department of Zoology, University of Cambridge, Cambridge, CB2 BEJ, England

compound is a phosphatidylglycerol, then X is:

$$CH_2OH$$
$$|$$
$$H \blacktriangleright C \blacktriangleleft OH$$
$$|$$
$$— CH_2$$

If the compound is a diphosphatidylglycerol (R_3 and R_4 being long chain acyl- groups), then X is:

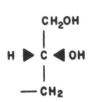

Other compounds of this type which are of interest are the polyphosphoinositides in which phosphatidyl-myo-inositol is phosphorylated on the 4, or the 4 and 5 positions of the inositol residue and also the phosphonolipids in which there is a direct phosphorus to carbon linkage between the glycerophosphate residue and the base.

The phosphosphingolipids (the second group) have the general formula:

$$CH_3(CH_2)_n \ CH = CH — \overset{H}{\underset{OH}{C}} — \overset{H}{\underset{NHCO—R}{C}} — CH_2—O—X$$

where R-CO is a long chain acyl group and X is a choline or ethanolamine residue.

CHEMICAL NATURE OF INSECT PHOSPHOLIPIDS

The comparative study of the phospholipids present in six orders of insect made by Fast (1966) has been extended to include an example from the primitive order Thysanura (Kinsella, 1969) and also the Isoptera (Mauldin, 1977). A considerable number of insect families have now been studied. It must be remembered that some

of the earlier work may not be so reliable as that of
more recent times with the more refined analytical techniques which are now available and the greater knowledge
of the subject. Because of the small size of insects
many of the analyses have been performed on extracts from
the whole animal--a situation not usually met with in
vertebrate work. This means that in some cases the un-
absorbed contents of the gut may influence the analysis.
It can be concluded that, in all insects for which a
reasonably full analysis of the different classes of
phospholipid has been performed, quantitatively the two
most important classes present are phosphatidylcholine
and phosphatidylethanolamine. In some cases--most of the
Diptera and Homoptera (Fast, 1966) and some of the Coleoptera (Kok and Norris, 1972a,b; Yadara and Musgrave,
1972)--the ethanolamine containing lipids are in clear
excess of the choline containing ones which is the reverse of that usually found in vertebrate tissue (White,
1973). In the case of the Diptera this observation may
be linked with order's inablility to synthesize choline
(Bridges, 1972). At least 80% of the total phospholipids
recovered from insects are of the choline and ethanolamine types. However, phosphatidyl-serine, -inositol and
-glycerol, phosphatidic acid and the polyglycerophosphatides have all been reported, although not always together in every insect. Monodeacylated derivatives, lyso-
compounds, of phosphatidyl-choline and -ethanolamine have
been shown to occur in varying proportion. The high
values reported in some instances may well be due to
breakdown of diacylphospholipids during or before extraction.

Sphingomyelin has been found in many insects but not
in the Diptera where its place appears to be taken by
ceramide phosphorylethanolamine, again possibly because
of the order's inability to synthesize choline (Bridges,
1972). This sphingolipid although initially found in insects (Phormia regina, Bieber, et al., 1963; Musca domestica, Crone abd Bridges, 1962; Calliphora erythrocephala,
Dawson and Kemp, 1968) has also been reported in shell-
fish, rumen protozoa and certain anaerobic bacteria
(Strickland, 1973). Both ceramide phosphorylethanolamine
and sphingomyelin occur in Apis mellifera (O'Connor et
al., 1970). The principal sphingosines present in insect
sphingolipids are tetradecasphing-4-enine and hexadeca-
sphing-4-enine (Musca domestica, Bieber et al., 1969a;
Apis mellifera; Acheta domesticus, Nauphoeta cineria and
Malacosoma americana, O'Connor, et al., 1970). This
finding differs from most vertebrate systems where C_{18}
compounds are the major sphingosines and is clearly not
linked to the presence of either ceramide phosphorylethanolamine or sphingomyelin.

The occurrence of alanine in the phospholipid fraction extracted from the tobacco hornworm, Manduca sexta has been reported (Hodgson, 1965; Kulkarni et al., 1971) along with another amino acid component, tentatively identified as leucine (Kulkarni et al., 1971). Another report refers to the presence of amino acids in the lipid fraction from Dendroctonus frontalis (Willis and Hodgson, 1970a). It is possible that some insects may contain phospholipids with small peptides linked to the more normal base components but considerably more work is necessary to confirm this. There is no evidence for the association of ^{14}C-alanine with the phospholipid fraction in Musca domestica larvae (Bridges and Price, 1970a).

When examined, it has been shown that diacylphospholipids are the major type present, most of the esterified fatty acids being 16 and 18 carbon atoms long with a high proportion of unsaturates (Fast, 1966). There have been only relatively few analyses which would have detected the presence of alk-1-enyl- or alkyl-groups. Where such analyses have been performed (Musca domestica, Crone and Bridges, 1963; Khan and Hodgson, 1967a; Periplaneta americana, Heliothis virescens and Anthonomus grandis Lambremont and Wood, 1968; Dendroctonus frontalis, Willis and Hodgson, 1970a) glycerylether phospholipids have been found in small amounts, representing only a minor proportion of the total phospholipid. However, most of the alk-1-enyl- and alkyl-groups are found in the ethanolamine phosphoglycerides of Dendroctonus frontalis (Willis and Hodgson, 1970a) and Heliothis virescens (Lambremont, 1972) and in these insects they cannot be considered to be minor constituents.

The presence of polyphosphoinositides has not been looked for widely in insects. Lipid fractions containing phosphorus and behaving chromatographically as triphospho- and diphospho-inositides have been extracted from whole larvae and from the larval nervous system of Musca domestica (Bridges, 1973a). Their presence in muscle and sensory tissue of Agrotis ypsilon and Acheta domesticus (Kilian and Schacht, 1979) and in the salivary gland of Calliphora erythrocephala (Fain and Berridge, 1979a) has been reported.

One of the more interesting aspects of insect phospholipids is the ease with which it is possible to effect an alteration in what are usually taken to be the normal phospholipid bases (choline and ethanolamine) by feeding analogues of the two bases to growing larvae. ß-Methylcholine and carnitine, which is decarboxylated by the insect to give the same analogue, are found to replace the growth requirement for choline by larvae of Phormia regina. The choline analogue is incorporated into phospholipid in place of what would normally be phosphatidylcholine (Bieber et al., 1963). N-Dimethylaminopropyl and N-dimethylaminoisopropyl alcohol are also found to be incorporated into phospholipids of the blowfly, the lipid formed from the former compound replacing some of the phosphatidylethanolamine as well as phosphatidylcholine (Bieber and Newburgh, 1963). A range of choline analogues which will act as choline replacements in larvae of Phormia regina and will become incorporated into phospholipid has been studied (Hodgson and Dauterman, 1964; Hodgson et al., 1969; Mehendale et al., 1970). A similar situation occurs with larvae of Musca domestica, in which a range of choline analogues are incorporated into phospholipids in place of choline when the analogues are included in the diet in the absence or with equivalent amounts of choline (Bridges et al., 1965; Bridges and Ricketts, 1970). In addition, when analogues of ethanolamine and their N-methyl-derivatives are included in the housefly larval diet, many are incorporated into the larva's phospholipids reducing the percentage of phosphatidylethanolamine normally present (Bridges and Ricketts, 1967). N-Dimethylaminoethanol replaces both ethanolamine and choline. The closer that the structure of the analogue is to either ethanolamine or choline then the more effectively is it utilised by the insect in place of the natural compounds. When suitable precursors are fed to housefly larvae the formation of phospholipids occurs. 2-Aminoethylphosphonate (Bridges and Ricketts, 1966), trimethyl- and dimethyl-aminoethylphosphonate (Bieber, 1968) are incorporated into the larval lipids, the first replacing phosphatidylethanolamine, the second phosphatidylcholine and the third both of the normal phospholipids.

Because of the inability of Diptera to synthesize choline, any choline analogue incorporated into the larval phospholipids will be carried over to the adult insect (provided a sufficient, small amount of choline is available to the larvae to satisfy the adult's need for synthesis of acetylcholine, Bridges, 1973b, 1974). This will remain as the phosphatidyl-derivative for the rest of the adult's lifetime. This is not so with ethanolamine analogues, as ethanolamine can be synthesised by all stages of the housefly and the percentage of the total phospholipid which is present as a phosphatidyl-analogue of ethanolamine is much reduced during the pupal period and continues to fall through the life span of the adult (Sohal and Bridges, 1977). However if adult flies are provided with glucose and solutions of ethanolamine- or choline-analogues, incorporation of the analogue into phospholipid can again occur (Bridges and Holden, 1969).

There is a small amount of evidence that some insects other than the Diptera can incorporate unnatural bases into their phospholipids. The most compelling is the report that ß-methylcholine (but not carnitine) will replace part of the choline in the larval lipids of

Tenebrio molitor (Bieber and Monroe, 1969). The partial
replacement of the choline requirement of Argyrotaenia
velutinana and Heliothis virescens by choline analogues
has been reported but the formation of unnatural phospholipids by these insects was not examined (Hodgson et al.,
1977).   Larvae of Pieris brassicae although requiring
choline in the diet for maximal growth, synthesize most
of their choline requirement and will replace approx. one
half of their normal phosphatidylcholine with phosphati-
dyl-N-dimethylethyl-choline and approx. one quarter with
phosphatidyl-ß-methylcholine when these analogues are included in the larval diet in place of choline (Bridges,
unpublished observations).   However these insects seemed
unable to incorporate analogues of ethanolamine into
their phospholipids (Bridges, unpublished observations).

Thus, although it is usually concluded that phospholipids of insects have the normal base composition it is
possible that, because the larval diet may contain an unusual foodstuff, phospholipids of ethanolamine or choline
analogues might be present.   So far, the changes that can
be induced in phospholipid bases are confined to ethanolamine and choline.   Analogues of serine are not found in
the lipid extracts from housefly larvae when such compounds have been included in the larval diet (Bridges and
Ricketts, 1967).   Six of the possible eight additional
isomers of inositol that are currently available for
testing will not substitute for the myo-inositol requirement of Pieris brassicae for growth and are not incorporated into larval lipids (Bridges, unpublished observations).

PHOSPHOLIPIDS AND GROWTH METAMORPHOSIS

A number of surveys of the phospholipid content and
type have been made covering the lifetime of the insect
from egg to adult.   These include studies made on Tribolium confusum (Beaudoin et al.,   1968), Phormia regina
(Bieber et al., 1961), Ceratitis capitata, Dacus oleae
(Castillon et al., 1971), Lucilia cuprina (D'Costa and
Birt, 1966), Anthonomus grandis (Henson et al.,   1972),
Culex pipiens fatigans (Kalra et al.,   1969), Acheta
domesticus (Lipsitz and McFarlane, 1970), Vitula edmundsae serratilineela (Miller and Blankenship, 1973), Trogoderma granarium (Rao and Agarwal, 1971), Diatraea grandiosella (Thompson et al., 1973a).   None of these show
any major changes in the proportion of the different
phospholipids present at each stage.   The position can be
summarised as follows: the major synthesis of phospholipids de novo occurs during the growing period of the
larval stage: the phospholipid content follows a U-shape
curve during pupation with a fall during histolysis followed by a rise as the pharate adult develops; if the

adult is supplied with sugar and water, there is only a slight fall in the phospholipid content during the lifetime of the adult (Bridges and Sohal, 1980). The phospholipid content of the ootheca of Periplaneta americana increases four-fold during development (Kinsella, 1966) whereas the phospholipid content of Plodia unipunctella eggs shows little change on maturation (Yurkiewicz, 1967) and the content actually decreases in the maturing eggs of Spodoptera exigua (Hoppe et al., 1975). Thus there seems to be no general response in phospholipid synthesis to egg development and it is unlikely that phospholipids act as any major reserve of energy to be used by the insect during growth and metamorphosis.

THE PHOSPHOLIPIDS OF DIFFERENT TISSUE AND IN SUBCELLULAR FRACTIONS

Where a particular tissue has been isolated from various insects and the phospholipid types present in the lipid extract of it determined, there have been few results to suggest that the phospholipid pattern of the tissue differs greatly from that of the whole insect. For example, fat body tissue from Diptera (Sarcophaga bullata, Allen and Newburgh, 1965; Crone et al., 1966; and Musca domestica, Bridges and Price, 1970b) had phosphatidylethanolamine as the predominant phospholipid whereas the same tissue from a lepidopteran, Manduca sexta, had phosphatidylcholine as the major phospholipid (Willis and Hodgson, 1970a). Analysis of the phospholipids present in a series of different tissues from housefly larvae including nervous tissues, muscle, cuticle, gut and fat body (Bridges and Price, 1970a,b) all show what is essentially the same distribution of the different phospholipid types in all tissues. The only major exceptions are when larvae are reared on a choline deficient diet and on a diet containing 2-aminobutan-1-ol. In the former case some preferential uptake of choline into the nervous system is observed and the phosphatidylcholine, when expressed as a percentage of the total tissue phospholipid, is three to five times greater than in other tissue (Bridges and Price, 1970a). The percentage of phosphatidyl-2-amino-butan-1-ol in the tissue phospholipid is not uniform in the different tissues. A much higher figure is found in the fat body than in other tissues suggesting that this tissue might be the major site of synthesis for phosphatidyl-2-amino-butan-1-ol in the fly.

Because of the function of the haemolymph lipoproteins as carriers of lipids and hormones from one tissue to another, the phospholipids of the haemolymph have been studied widely. If the haemolymph is centrifuged to remove haematocytes the remainder of the phosphlipids is

associated with lipoproteins. In housefly larvae the
phospholipid composition of the unfractionated haemolymph (Bridges and Price, 1970b) and the lipoproteins
that can be separated from it (Dwivedy and Bridges, 1973)
is very different from any other tissue of the insect.
A much smaller proportion of polyglycerophosphatides and
phosphatidylcholine with a corresponding larger percentage of phosphatidylethanolamine is observed. Phosphatidylethanolamine is reported to be the principal phospholipid in Acheta domesticus haemolymph (Wang and Patton,
1969) and in the high density lipoprotein isolated from
Manduca sexta larval haemolymph (Pattnaik et al., 1979)
but in other insects examined the major phospholipid in
the haemolymph or haemolymph lipoprotein is phosphatidylcholine (Schistocerca gregaria, Mehrotra et al., 1966;
Galleria mellonella, Wlodawer and Wisniewska, 1965; Locusta migratoria migratoriodes, Peled and Tietz, 1975;
Manduca sexta, Mundell and Law, 1979; pupal Hyalophora
cecropia, Thomas and Gilbert, 1968; and pupal Philosamia
cynthia, Chino et al., 1969). A recent review of animal
lipoproteins (Chapman, 1980) points out the enrichment
of phosphatidylethanolamine in the phospholipids of insect haemolymph lipoproteins compared with vertebrate
lipoproteins. This enrichment is a feature of insects
irrespective of whether their major phospholipid in other
tissues is phosphatidylethanolamine or phosphatidylcholine. Whether these lipoproteins are involved in the
transport of phospholipids between the different insect
tissues as has been suggested (Thomas and Gilbert, 1968;
Katayiri and Chino, 1973) or whether specific proteins
are involved in their extra- and intra-cellular transport
as has been shown to be the case in vertebrates (Wirtz,
1974) remains to be seen. The phospholipid composition
of subcellular fractions from whole insect or insect
tissue has been determined. The mitochondrial (or sar-
cosomal, for adult flight muscle) fraction from various
Diptera (Musca domestica, Bridges and Price, 1970b, Khan
and Hodgson, 1967a, Crone, 1964, Chan, 1970, Sohal and
Bridges, 1977; Glossina morsitans, D'Costa and Rutesa-
sina, 1973) shows a slightly different distribution of
phospholipid types from that of whole insect or muscle.
A higher percentage of polyglycerophosphatide and lower
percentage of phosphatidylcholine is a fairly common feature of the results. Mitochondrial preparations from the
muscle of adult Schistocerca gregaria (Novakova et al.,
1976) and pharate pupae of Anthonomus grandis (Thompson
et al., 1973b) have similar percentage values of phosphatidylcholine and phosphatidylethanolamine; here again the
preparations have a higher percentage of polyglycerophosphatides than do other fractions. Mitochondria from
white muscle of Schistocerca gregaria have a rather different composition from those from red muscle (Novakova

et al., 1976). A continuous increase in the percentage of polyglycerophosphatides occurs at the expense of the phosphatidylcholine with development of mitochondria in pupae during the five days before the emergence of the adult Manduca sexta (Chan and Lester, 1970). The higher percentage value of polyglycerophosphatides found in the mitochondrial fraction means that the insect preparations are similar to those of vertebrates where this phospholipid is largely associated with the inner mitochondrial membrane (McMurray, 1973). However the value (24.6% of the mitochondrial phospholipid) obtained by Chan (1970) for cardiolipin in the mitochondrial fractions from Musca domestica is considerably higher than that found by other workers using this or other insect preparations.

Analyses have been performed on microsomal fractions prepared from some insects (Musca domestica, Bridges and Price, 1970b, Khan and Hodgson, 1967a; Anthonomus grandis, Thompson et al., 1973b). It has been shown in a study on different tissue sub-cellular fractions of housefly larvae containing an unnatural phospholipid (Bridges and Price, 1970b) that all fractions contain a similar percentage of the unnatural lipid in the total phospholipid of the fraction. There is no suggestion that any one particular subcellular fraction of gut, muscle or fat body tissue has any special requirement for one type of phospholipid (apart from the mitochondria for polyglycerophosphatide mentioned above).

SYNTHESIS OF PHOSPHOLIPIDS

Examination of the routes of synthesis of phospholipids in insects has been sporadic and incomplete. All the evidence to date suggests that insects do not utilize any novel pathways for phospholipid synthesis. Evidence for the cytidine pathway of phosphatidyl-choline and -ethanolamine synthesis in some insects is good. Four enzymes are involved, initially a kinase for the production of the phosphorylbase, then a cytidyltransferase to form the cytidine diphosphorylbase and then a diglyceride base-phosphotransferase. The diglyceride is derived from phosphatidic acid by a phosphohydrolase. The isolation and partial purification of a choline kinase from the gut tissue of Periplaneta americana has been achieved (Kumar and Hodgson, 1970; Shelley and Hodgson, 1970; 1971a; Habibulla and Newburgh, 1969). Radioactive phosphate and phosphorylcholine are incorporated into the phospholipids of Celerio euphorbiae and Arctia caia (Chojnacki and Piechowska, 1961; Chojnacki and Korzybski, 1962). Incorporation of ^{32}P-labelled adenosinediphosphorylcholine, cytidinediphosphorylcholine or deoxycytidinediphosphorylcholine into phosphatidylcholine of fat body and brain preparations of Locusta migratoria and Bombyx mori shows that the adenosine derivative is inactive and the deoxycytidine derivative only one half to one third as effective as the cytidine compound (Chojnacki and Korzybski, 1963). ATP stimulates the incorporation of ^{14}C-choline into phosphorylcholine and phosphatidyl-choline and CTP stimulated the incorporation of phospho-rylcholine into phosphatidylcholine in cell free prepara-tion of the fat body of Sarcophaga bullata (Rao and Agar-wal, 1971). As the radioactivity from ^{32}P-phosphoryl-choline is incorporated into phosphatidylcholine without the appearance of the label in other lipids or water sol-uble compounds it is concluded that the route of synthe-sis is the same in this dipteran as known in vertebrates (Crone et al., 1966). Analogues of ethanolamine which are incorporated into phospholipid of housefly larvae and adults also appear to follow the same synthetic pathway, as both the phosphoryl- and CDP-derivatives of the amino alcohols are detected (Bridges and Ricketts, 1967, 1968; 1970; Bridges and Holden, 1969). A build up of the phosphoryl-derivatives suggests that the cytidylphospho-ryltransferase is the most specific of the three enzymes involved. The specificity of the phosphokinase for ethanolamine and choline in fat body preparations of Phormia regina suggests that there are in fact two phos-phokinases (or two molecular species) one of which is specific for choline and the other of which will phos-phorylate both choline and ethanolamine (Shelley and Hodgson, 1971a). Using the same system a number of cho-line analogues are effective competitors of choline phos-phorylation (Shelley and Hodgson, 1971b). Further evi-dence is obtained, again using fat body preparations of Phormia regina, for two separate kinases, one for etha-nolamine and one for choline, both of which are inhibited by certain aminoalcohols (Shelley and Hodgson, 1971b). β-Methylcholine is phosphorylated by the kinase from a similar preparation (Habibulla and Newburgh, 1969) and one from houseflies (Bieber et al., 1969b). Further evi-dence that Phormia regina larvae can incorporate choline analogues into phospholipids by the cytidine pathway has been obtained (Mehendale et al., 1970) along with the build up of the phosphate ester of the analogues. The presence of two kinases phosphorylating choline and etha-nolamine has been found in whole adult Culex fatigans (Ramabrahmam and Subrahmanyam, 1979).

The presence of phosphatidate phosphohydrolase in the fat body and muscle of Hyalophora cecropia demonstrates that the necessary enzyme for the supply of diglyceride from phosphatidic acid is present in at least one insect (Hirano and Gilbert, 1967).

Synthesis of the acidic phospholipids, phosphatidyl-inositol, phosphatidylserine and cardiolipin does not seem to have been studied in insects. Incorporation of

^{32}P from labelled phosphate has been demonstrated into such lipids in vivo and in vitro but the exact route of synthesis has not been established. The possibility that one phospholipid is formed from another by a base exchange reaction has been investigated in homogenate of housefly larval fat body (Crone, 1967a). Ethanolamine and serine act as competitive inhibitors with one another in an exchange which is stimulated by calcium. A range of other amino alcohols, N-monomethylaminoethanol, N-dimethylaminoethanol and choline all inhibit the exchange of ethanolamine into the preparation. The decarboxylation of phosphatidylserine to phosphatidylethanolamine occurs in housefly larvae of Musca domestica (Crone, 1967b), Phormia regina (Taylor and Hodgson, 1965) and Acantholyda nemoralis (Zielinska and Dominas, 1967). It is possible, therefore, that during the growth period of larvae when synthesis of phosphatidylethanolamine is required, phosphatidylserine may be decarboxylated and the phosphatidylethanolamine converted back to phosphatidylserine by base exchange in a cyclical manner. The released ethanolamine could be incorporated into phosphatidylethanolamine by the cytidine pathway leading to a net synthesis.

The direct methylation pathway of phosphatidylethanolamine via the phospholipid derivatives of mono- and dimethylaminoethanol is obviously not present in Diptera as the absence of any choline synthesis has been clearly established (Phormia regina, Bieber and Newburgh, 1963; Hodgson et al., 1960; Musca domestica, Bridges et al., 1965; Moulton et al., 1970). The utilization of N-monoethylaminoethanol and N-dimethylaminoethanol as substitutes for ethanolamine and choline by larvae of the housefly results in the formation of the corresponding phosphatidyl-derivative with no increase in phosphatidylcholine or in phosphatidyl-N-dimethylaminoethanol when N-monomethylaminoethanol is fed (Bridges and Ricketts, 1967). In fact, rather than being methylated, N-dimethylaminoethanol is dealkylated with a formation of phosphatidyl-N-monomethylaminoethanol. The presence or absence of the direct methylation pathway in other insects is less certain. Its absence in Tribolium confusum seems likely (Beaudoin and Lemonde, 1970) and there is some evidence that [$^{14}CH_3$] methionine does not give rise to radioactive choline in the lipid fraction extracted from Heliothis zea (Willis and Hodgson, 1970b). There is however no evidence that a choline can be synthesised by Tenebrio molitor larvae (Bieber and Monroe, 1969) and Vitula edmundsae serratilineela larvae (Blankenship and Miller, 1971). Larvae of Pieris brassicae have a requirement for only a small amount of choline for optimal growth and on a choline-free diet are able to synthesise phosphatidylcholine, implying the existence

of a methylation pathway from phosphatidylethanolamine
(Bridges, unpublished observations).

HYDROLYSIS OF PHOSPHOLIPIDS

The removal of the first two fatty acids esterified
on the 1 and 2 positions of the phosphoglycerides is
achieved by enzymes known as phospholipase A_1 and A_2
respectively. The remaining fatty acid is removed by a
lysophospholipase (phospholipase B). The presence of a
phospholipase A, stimulated by Ca^{2+}, which hydrolyses
phosphatidylethanolamine can be demonstrated in incubates
of homogenised fat bodies of larval and adult Musca domestica (Crone, 1967a). A similar hydrolysis of phosphatidyl-choline and -ethanolamine is shown to occur in
various subcellular particles of Musca domestica and
Phormia regina although largely concentrated in the microsomal and supernatant fractions (Khan and Hodgson,
1967b). A lysophospholipase (phospholipase B) hydrolys-
ing lysophosphatidyl-choline and -ethanolamine is more
concentrated in the mitochondrial fraction although there
may be two phospholipases involved as Ca^{2+} has a different effect on the hydrolysis of each lyso-lipid.
These enzymes have similar properties to those which have
been studied from vertebrate sources. Phospholipase A
and lyso-phospholipase activity have been demonstrated
in larval homogenates of Culex fatigans (Rao and Subrahmanyam 1969), in pupal and adult Glossina morsitans
(Isharaza et al., 1978) and the fat body of the larvae
of Bombyx mori (Yanagawa and Horie, 1978). The suggestion that glycerylphosphorylcholine in the haemolymph of
Hyalophora cecropia is a degradation product of phosphatidylcholine implied the presence of phospholipase A and
lyso-phospholipase in this insect (Carey and Wyatt, 1963)
as do the results obtained with Periplaneta americana
(Kumar and Hodgson, 1970). Microsomal preparations from
larvae of Musca domestica convert exogenous phosphatidyl-choline, -ethanolamine and -ß-methylcholine to the
respective glycerylphosphorylderivatives via the lyso-
derivatives (Kumar et al., 1970). A similar finding has
been reported with the acidic phospholipids, phosphatidyl-inositol, -serine and -glycerol (Hildenbrandt et al.,
1971). The complete hydrolysis of phosphatidyl-ß-methyl-
choline to α-glycerophosphate and ß-methylcholine has
been demonstrated in housefly larvae (Bieber et al.,
1969). Acyl-transferase activity is present in microsomal preparations of Musca domestica larvae which will
acylate 1-acylglycero-3-phospho-choline, -ethanolamine
and -ß-methylcholine (Kumar et al., 1970). There is
little evidence for the presence in insects of phospholipase C and D activities which hydrolyse glycerophospho-
lipids to diglyceride and the phosphorylbase or to phos-

phatidic acid and base respectively. Their absence has been noted in <u>Glossina morsitans</u> (Isharaza <u>et al</u>., 1978) and no evidence is found for phospholipase C activity in larval homogenates of <u>Musca domestica</u> (Hildebrandt <u>et al</u>., 1971). However the possible presence of a phospholipase C type of enzyme has been suggested to explain the water-soluble metabolites containing inositol formed from phosphatidylinositol in salivary glands of <u>Calliphora erythrocephala</u> (Fain and Berridge, 1979a), but no evidence for such activity is found in a microsomal preparation of <u>Musca domestica</u> larvae (Hildenbrandt <u>et al</u>., 1971). The presence of ceramide phosphorylethanolamine hydrolase activity along with ceramidase activity in microsomal preparation of <u>Musca domestica</u> with the production of ceramide, phosphorylethanolamine, sphingosine and fatty acid suggests that the metabolism of ceramide phosphorylethanomine follows the conventional course of sphingomyelin metabolism in this insect (Hildenbrandt <u>et al</u>., 1971).

THE ROLE OF PHOSPHOLIPIDS IN INSECTS

The correlation between the insect's phospholipid content and weight during growth and the resistance of this content to change during periods of starvation suggest that, as in vertebrates, the main role of phospholipids is as a component of membranes and not as an energy reserve. This is again borne out by the presence of phospholipids in all the membrane subcellular fractions examined from insects. However the need for one or more class of phospholipid in particular amount or proportion in the make up of the membranes is less certain. Almost all Diptera have a much higher proportion of phosphatidylethanolamine and lower proportion of phosphatidylcholine than many other insects and vertebrates and in <u>Musca domestica</u> much of this already small content of phosphatidylcholine can be replaced by phosphatidylethanolamine. The only need for choline would seem to be to supply sufficient precursor for acetylcholine synthesis and once this is satisfied there is no need for choline for synthesis of further phosphatidylcholine (Bridges, 1972). At this stage choline and ethanolamine lipids appear to be interchangeable and the base components of the two major phospholipids of the housefly appear to be of no differential benefit to the insect. In fact analogues may be incorporated into the lipids in the place of the normal bases. A change of phospholipid base might be expected to be accompanied by a change in the fluidity of the membrane bilayer. It has been suggested that the high levels of palmitoleic acid in phospholipids of the Diptera might have a compensating effect for the presence of the high levels of phosphatidyletha-

nolamine by maintaining the normal membrane fluidity
(Fast, 1966). However other insects have a high proportion of phosphatidylethanolamine in their lipids without
any unusual distribution of esterified fatty acids (Yadava et al., 1972). Also, although changes in the fatty
acids esterified to the phospholipids of housefly larvae
were observed when different levels of phosphatidylethanolamine were induced by dietary means the changes were
not those predicted to maintain normal fluidity (Bridges
and Watts, 1975). The high proportion of unsaturated
acids in insect phospholipids may well mean that the head
group has very little effect on membrane fluidity. An
increase in the proportion of esterified saturated fatty
acids has been said to occur with decreasing rearing
temperature in order to maintain the fluidity of the insects' membranes. A recent example of this is the report
that temperature dependent changes in fatty acids composition were observed in housefly larval glycerophospho-
lipids together with a change in the sphingosine residue
of the ceramide phosphorylethanolamine (Robb et al 1972).
However other workers could not reproduce these findings
concerning the fatty acids of the glycerophosphatides
(Bridges and Watts, 1975). The possibility that change
in phospholipid head group with resulting change in fluidity could be compensated for by change in proportion
of cholesterol in the various tissues of the housefly was
examined and concluded to be unlikely (Dwivedy, 1979).

A specific requirement for acidic phospholipids may
be necessary. Phosphatidylserine could not be replaced
by phospholipids containing analogues of serine in housefly larvae (Bridges and Ricketts, 1967). This finding
is perhaps not very convincing as serine, unlike choline,
is readily synthesised by the fly. The meaning of the
atypical incorporation of ^{32}P-phosphate into phosphatidylinositol compared to other phospholipids in muscle
of Periplaneta americana and Schistocerca gregaria (Novakova et al., 1976; Helm et al., 1977; Strunecka et al.,
1978) is not clear, but suggests that there may be a
special role for this phospholipid. A requirement for
myo-inositol in the larval diet of Pieris brassicae could
not be replaced by other isomers (Bridges, unpublished
observations). This specific requirement for phosphatidyl-myo-inositol might be related to the possible connection between the turnover of this lipid and calcium gat-
ing which has been investigated in depth using salivary
glands from adult Calliphora erythrocephala (Fain and
Berridge, 1979a,b, Berridge and Fain, 1979). These
studies on the hormonal regulation of phosphatidylinositol turnover appear to be facilitated by several unusual
features of the insect salivary gland. The possibility
that prostaglandins synthesised from arachidonic acid
released from phosphatidylinositol, are involved as a

second messenger to release bound calcium (Barritt, 1981)
seems extremely unlikely in the housefly which can be
reared on diets containing no fatty acids. Such insects
synthesise only saturated and mono-unsaturated fatty
acids to give fully developed insects containing no detectable esterified polyunsaturated acids (Bridges and
Sohal, 1980). The latter observation also suggests that
prostaglandins are not utilised by the larval or adult
housefly, except possibly for reproduction. The finding
that pulsed tones increase the incorporation of ^{32}P-
phosphate into both monophospho- and diphosphophospha-
tidylinositol but not into adenosine triphosphate, phosphatidyl-inositol or -serine and phosphatidic acid in the
auditory organ of Agrotis ypsilon suggests that polyphosphoinositides may play a role in events relating to
neural excitation in insects (Kilian and Schacht, 1980).
The absence of methyltransferases for the synthesis
of phosphatidylcholine from phosphatidylethanolamine in
Diptera makes the proposed utilisation of this pathway
as a means of controlling ion channels (Hirata and Axelrod, 1980) in these insects not feasible.
In summary, therefore, no new role for phospholipids
has been found in insects. However the very simplicity
of the metabolic systems used by the insect makes it
likely that the more exotic suggestions proposed for the
involvement of phospholipids in vertebrate cell metabolism can at their best be no more than modifiers of the
basic and fundamental processes common to the animal
kingdom.

REFERENCES

Allen, R.R. and Newburgh, R.W. (1965) Phospholipid composition of fat bodies of Sarcophaga bullata. J.
Insect Physiol. 11: 1601-1603.
Barritt, G.J. (1981) A proposal for the mechanism by
which ß-adrenergic agonists, vasopressin angiotensin
and cyclic AMP induce calcium release from intracellular stores in the liver cells: a possible role for
metabolites of arachidonic acid. Cell Calcium 2:
53-63.
Beaudoin, A.R. and Lemonde, A. (1970) Aspects du metab-
olisme des phospholipides des Tribolium confusum. J.
Insect Physiol. 16: 511-519.
Beaudoin, A.R., Villeneuve, I.-L., and Lemonde, A. (1968)
Variation des phospholipides au cours de la métamor-
phose de Tribolium confusum (Coléoptère). J. Insect
Physiol. 14: 831-840.
Berridge, M.J. and Fain, J.N. (1979) Inhibition of
phosphatidylinositol synthesis and the inactivation
of calcium entry after prolonged exposure of the
blowfly salivary gland to 5-hydroxytryptamine. Bio-

chem. J. 178: 56-69.

Bieber, L.L. (1968) Incorporation of trimethylamino-ethylphosphonic acid and dimethylaminoethylphosphonic acid into lipids of housefly larvae. Biochim. biophys. Acta 152: 778-780.

Bieber, L.L. and Monroe, R.E. (1969) The relation of carnitine to the formation of phosphatidyl-ß-methyl-choline by Tenebrio molitor L. larvae. Lipids 4: 293-298.

Bieber, L.L. and Newburgh, R.W. (1963) The incorporation of dimethylaminoethanol and dimethylaminoisopropyl alcohol into Phormia regina phospholipids. J. Lipid Res. 4: 397-401.

Bieber, L.L., Hodgson, E., Cheldelin, V.H., Brookes, V.J. and Newburgh, R.H. (1961) Phospholipid patterns in the blowfly Phormia regina (Meigen). J. biol. Chem. 236: 2590-2595.

Bieber, L.L., Cheldelin, V.H., and Newburgh, R.W. (1963) Studies on a ß-methylcholine-containing phospholipid derived from carnitine. J. biol. Chem. 4: 1262-1265.

Bieber, L.L., O'Connor, J.D. and Sweeley, C.C. (1969a) The occurrence of tetradecasphing-4-enine and hexadecasaphing-4-enine as the principal sphinogosines of Musca domestica larvae and adults. Biochim. biophys. Acta 187: 157-159.

Bieber, L. L., Sellers, L. G., and Kumar, S. S. (1969b) Studies on the formation and degradation of phospha-tidyl-ß-methylcholine, ß-methylcholine derivatives and carnitine by the housefly larvae. J. biol. Chem. 244: 630-636.

Blankenship, J.W. and Miller, G.J. (1971) In vivo synthesis of phosphatidylcholine and ethanolamine in larvae of the dried fruit moth. J. Insect Physiol. 17: 2061-2067.

Bridges, R.G. (1972) Choline metabolism in insects. Adv. Insect. Physiol. 9: 51-110.

Bridges, R.G. (1973a) The lipid composition of the larval nervous system of Musca domestica. A comparison between insects susceptible and resistant to cylodiene insecticides. Comp. Biochem. Physiol. 44B: 191-203.

Bridges, R.G. (1973b) Preferential incorporation of choline into the lipids of the nervous system of the housefly Musca domestica. J. Insect Physiol. 19: 2439-2443.

Bridges, R.G. (1974) Acetylcholine in the choline deficient housefly, Musca domestica. J. Insect Physiol. 20: 2363-2374.

Bridges, R.G. and Holden, J.S. (1969) The incorporation of aminoalcohols into the phospholipids of the adult housefly, Musca domestica. J. Insect Physiol. 15: 779-788.

Bridges, R.G. and Price, G.P. (1970a) Phospholipid composition of various organs from larvae of the housefly, Musca domestica, fed on diets deficient in choline. Int. J. Biochem. 1: 483-490.

Bridges, R.G. and Price, G.M. (1970b) The phospholipid composition of various organs from larvae of the housefly, Musca domestica, fed on normal diets and on diets containing 2-aminobutan-1-ol. Comp. Biochem. Physiol. 34: 47-60.

Bridges, R.G. and Ricketts, J. (1966) Formation of a phosphonolipid by larvae of the housefly, Musca domestica. Nature, Lond. 211: 199-200.

Bridges, R.G. and Ricketts, J. (1967) The incorporation, in vivo, of aminoalcohols into the phospholipids of the larvae of the housefly, Musca domestica. J. Insect Physiol. 13: 835-850.

Bridges, R.G. and Ricketts, J. (1968) The effect of 2-amino-butan-1-ol on the growth of the housefly (Musca domestica). Comp. Biochem. Physiol. 25: 383-400.

Bridges, R.G. and Ricketts, J. (1970) The incorporation of analogues of choline into the phospholipids of the housefly, Musca domestica. J. Insect Physiol. 16: 579-593.

Bridges, R.G., Ricketts, J. and Cox, J.T. (1965) The replacement of lipid-bound choline by other bases in the phospholipid of the housefly, Musca domestica. J. Insect Physiol. 11: 225-236.

Bridges, R.G. and Sohal, R.S. (1980) Relationship between age-associated fluorescence and linoleic acid in the housefly, Musca domestica. Insect Biochem. 10: 557-562.

Bridges, R.G. and Watts, S.G. (1975) Changes in fatty acid composition of phospholipids and triglycerides of Musca domestica resulting from choline deficiency. J. Insect Physiol. 21: 861-871.

Carey, F.G. and Wyatt, G.R. (1963) Phosphate compounds in tissues of Cecropia silk moth during diapause and development. J. Insect Physiol. 9: 317-335.

Castillon, M.P., Catalan, R.E., Municio, A.M. and Suarez, A. (1971) Biochemistry of the development of the insects Dacus oleae and Ceratitis capitata: evolution of the phospholipids. Comp. Biochem. Biophys. 38B: 109-117.

Chan, S.K. (1970) Phospholipid composition in the mitochondria of the housefly Musca domestica. A reexamination. J. Insect Physiol. 16: 1575-1577.

Chan, S.K. and Lester, R.L. (1970) Biochemical studies on the developing thoracic muscles of the tobacco hornworm. II. Phospholipid of the mitochondria during development. Biochim. biophys. Acta 210: 180-181.

Chapman, M.J. (1980) Animal lipoproteins: chemistry,

structure and comparative aspects (review). J. Lipid. Res. 21: 789-853.

Chino, H., Murkahami, S. and Harashima, K. (1969) Diglyceride carrying lipoproteins in insect haemolymph: isolation purification and properties from pupal haemolymph. Biochim. biophys. Acta. 176: 1-26.

Chojnacki, T. and Korzybski, T. (1962) Biosynthesis of phospholipids in insects. III The incorporation of ^{32}P orthophosphate into phospholipids of Artica caia moths. Acta. Biochim. Polon. 9: 95-110.

Chojnacki, T. and Korzybski, T. (1963) On the specificity of cytidine coenzyme in the incorporation of phosphorylcholine into phospholipids by tissue homogenates of various animal species. Acta. Biochim. Polon. 10: 455-461.

Chojnacki, T. and Piechowska, M.J. (1961) Biosynthesis of phospholipids in insects II. Incorporation of [^{32}P] phosphocholine into phospholipids of Celerio euphorbia. Acta Biochim. Polon. 8: 157-165.

Crone, H.D. (1964) Phospholipid composition of flight muscle sarcosomes from the housefly, Musca domestica. J. Insect Physiol. 10: 499-507.

Crone, H.D. (1967a) The calcium-stimulated incorporation of ethanolamine and serine into phospholipids of the housefly, Musca domestica. Biochim. J. 104: 695-704.

Crone, H.D. (1967b) The relationship between phosphatide serine and ethanolamine in the larvae of the housefly, Musca domestica. J. Insect Physiol. 11: 81-90.

Crone, H.D. and Bridges, R.G. (1962) The phospholipids of the housefly (Musca domestica) stable to hydrolysis by mild alkali and acid. Biochem. J. 84: 101.

Crone, H.D. and Bridges, R.G. (1963) The phospholipids of the housefly, Musca domestica. Biochem. J. 89: 11-21.

Crone, H.D., Newburgh, R.W. and Mezei, C. (1966) The larval fat body of Sarcophaga bullata (Diptera) as a system for studying phospholipid biosynthesis. J. Insect Physiol. 12: 619-624.

Dawson, R.M.C. and Kemp, P. (1968) Isolation of ceramide phosphorylethanolamine from the blowfly Calliphora erythrocephala. Biochem. J. 106: 319-320.

D'Costa, M.A. and Birt, L.M. (1966) Changes in the lipid content during metamorphosis of the blowfly Lucilia cuprina. J. Insect Physiol. 12: 377-394.

D'Costa, M.A. and Rutesasira, A. (1973) Phospholipid composition of flight muscle sarcosome from the tsetse fly, Glossina morsitans. Comp. Biochem. Physiol. 43B: 491-498.

Dwivedy, A.K. (1979) Effect of altered phospholipid head groups on the distribution of cholesterol in the housefly (Musca domestica) larvae. Insect Biochem.

9: 273-278.

Dwivedy, A.K. and Bridges, R.G. (1973) The effect of dietary changes on the phospholipid composition of the haemolymph lipoproteins of larvae of the housefly, Musca domestica. J. Insect Physiol. 19: 559-576.

Fain, J.N. and Berridge, M.J. (1979a) Relationship between hormonal activation of phosphatidylinositol hydrolysis, fluid secretion and calcium flux in the blowfly salivary gland. Biochem. J. 178: 45-58.

Fain, J.N. and Berridge, M.J. (1979b) Relationship of phosphatidylinositol synthesis and recovery of 5-hydroxytryptamine responsive calcium flux in blowfly salivary glands. Biochem. J. 180: 655-661.

Fast, P.G. (1966) A comparative study of the phospholipids and fatty acids of some insects. Lipids 1, 209-215.

Habibulla, M. and Newburgh, R.W. (1969) Carnitine de carboxylase and phosphokinase in Phormia regina. J. Insect Physiol. 15: 2245-2253.

Helm, R., Novak, F., Sula, J., Novakova, O., and Kubista, V. (1977) Phospholipid metabolism in the flight muscle of Periplaneta americana during maturation. Insect Biochem. 7: 73-76.

Henson, R.D., Thompson, A.C., Gueldner, R.C. and Hedin, P.A. (1972) Variations in lipid content of the boll weevil during metamorphosis. J. Insect Physiol. 18: 161-167.

Hildenbrandt, G.R., Abraham, T. and Bieber, L.L. (1971) Metabolism of ceramide phosphorylethanolamine, phosphatidylinositol, phosphatidylserine and phosphatidylglycerol by housefly larvae. Lipids 6: 508-516.

Hirano, C. and Gilbert, L.I. (1967) Phosphidate phosphohydrolase in the fat body of Hyalophora cecropia. J. Insect Physiol. 13: 163-174.

Hirata, F. and Axelrod, J. (1980) Phospholipid methylation and biological signal transmission. Science 209: 1082-1090.

Hodgson, E. (1965) Phospholipids of the tobacco hornworm Protoparce sexta Johan (Lepidoptera Sphingidae). Experimentia 21: 78-79.

Hodgson, E. and Dauterman, W.C. (1964) The nutrition of choline, carnitine and related compounds in the blowfly, Phormia regina Meigen. J. Insect Physiol. 10: 1005-1008.

Hodgson, E., Cheldelin, V.H. and Newburgh, R.W. (1960) Nutrition and metabolism of methyl donors and related compounds in the blowfly, Phormia regina (Meigen). Arch. Biochem. Biophys. 87: 48-54.

Hodgson, E., Dauterman, W.C., Mehendale, H.M., Smith, E. and Khan, M.A.Q. (1969) Dietary choline requirements, phospholipids and development in Phormia

178

regina. Comp. Biochem. Biophys. 29: 343-359.

Hodgson, E., Ligon, B.G. and Rock, G.C. (1977) Substitution of choline by related compounds in the diets of Argyrotaenia velutinana and Heliothis virescens. J. Insect Physiol. 23: 801-804.

Hoppe, K.T., Hadley, N.F. and Trelease, R.N. (1975) Changes in lipid and fatty acid composition of eggs during development of the beet army worm, Spodoptera exigua. J. Insect Physiol. 21: 1427-1430.

Isharaza, W.K., Kakonge, E.J. and Lutalo-Boja, A.J. (1978) Phospholipases of the tsetse fly Glossina morsitans. Comp. Biochem. Physiol. 59B: 87-93.

Kalra, R.L., Wattal, B.L. and Venkitasubramanian, T.A. (1969) Lipids of Culex pipiens fatigans during metamorphosis. Indian J. Exp. Biol. 7: 154-157.

Katagiri, C. and Chino, H. (1973) Studies on phospholipid transport by haemolymph lipoproteins. Insect Biochem. 3: 429-437.

Khan, M.A.Q. and Hodgson, E. (1967a) Phospholipids of subcellular fractions from the houseflies, Musca domestica. J. Insect Physiol. 13, 645-664.

Khan, M.A.Q. and Hodgson, E. (1967b) Phospholipase activity in Musca domestica L. Comp. Biochem. Physiol. 23, 899-910.

Kilian, P.L. and Schacht, J. (1979) Polyphosphoinositides in insect muscle and sensory tissue. J. Neurochem. 32: 247-248.

Kilian, P.L. and Schacht, J. (1980) Sound stimulates labeling of polyphosphoinositides in the auditory organ of the noctuid moth. J. Neurochem. 34: 709-712.

Kinsella, J.E. (1966) Phospholipid patterns of Periplaneta americana during embryogenesis. Comp. Biochem. Biophys. 17: 635-640.

Kinsella, J.E. (1969) The lipids of Lepisma saccharina L. Lipids 4: 299-300.

Kok, L.T. and Norris, D.M. (1972a) Lipid composition of adult female Xyleborus ferrugineus. J. Insect Physiol. 18: 1137-1151.

Kok, L.T. and Norris, D.M. (1972b) Comparative phospholipid composition of adult female Xyleborus ferrugineus and its mutualistic fungal ectosymbionts. Comp. Biochem. Physiol. 42B: 245-254.

Kulkarni, A.P., Smith, E. and Hodgson, E. (1971) The phospholipids of Manduca sexta tissues and incorporation in vivo of ethanolamine, choline and inorganic phosphate. Insect Biochem. 1: 348-362.

Kumar, S.S. and Hodgson, E. (1970) Partial purification and properties of choline kinase from cockroach, Periplaneta americana. Comp. Biochem. Physiol. 33: 73-85.

Kumar, S.S., Millay, R.H. and Bieber, L.L. (1970) De-

acylation of phospholipids and acylation and deacyla-
tion of lysophospholipids containing ethanolamine
choline and ß-methylcholine by microsomes from housefly larvae. Biochemistry 9: 754-759.

Lambremont, E.N. (1972) Ether bonded lipids of insects.
A quantitative comparison of the glycerylethers
associated with ethanolamine and choline phosphoglycerides. Comp. Biocem. Physiol. 41B: 337-342.

Lambremont, E.N. and Wood, R. (1968) Glycerylethers in
insects, identification of alkyl and alk-1-enyl glycerylether phospholipids. Lipids 3: 503-510.

Lipsitz, E.Y. and McFarlane, J.E. (1970) Total lipid
and phospholipid during the life cycle of the house
cricket. Comp. Biochem. Biophys. 34: 699-705.

Mauldin, J.K. (1977) Cellulose catabolism and lipid synthesis by normally and abnormally faunated termites,
Reticulitermes flavipes. Insect Biochem. 7: 27-31.

McMurray, W.C. (1973) Phospholipids in subcellular
organelles and membranes. In "Form and Function of
Phospholipids" (eds. Ansell, G.B., Hawthorne, J.N.
and Dawson, R.M.C.). Elsevier Scientific Publishing
Company, Amsterdam-London-New York. 205-251.

Mehendale, H.M., Dauterman, W.C. and Hodgson, E. (1970)
The incorporation of choline analogues into the phospholipids of Phormia regina. Int. J. Biochem. 1:
429-437.

Mehrotra, K.N., Sethi, G.R. and Bhamburhar, M.W. (1966)
Phospholipids in the haemolymph of desert locusts
Schitocerca gregaria F. Indian J. Ent. 28: 468-476.

Miller, G.J. and Blankenship, J.W. (1973) Influence of
dietary lipids upon lipids in larvae and adults of
the dried fruit moth. J. Insect Physiol. 19: 65-74.

Moulton, B., Rottman, F., Kumar, S.S. and Bieber, L.L.
(1970) Utilization of methionine, choline and ß-me-
thyl choline by Musca domestica. Biochim. biophys.
Acta 210: 182-185.

Mundall, E.C. and Law, J.H. (1979) Physical and chemical characterisation of vitellogenin from the haemolymph and eggs of the tobacco hornworm, Manduca
sexta. Comp. Biochem. Physiol. 63B: 459-468.

Novaka, O., Novak, F. and Kubista, V. (1976) Phospholipid metabolism in red and white insect muscle.
Insect Biochem. 6: 381-384.

O'Connor, J. D., Polito, A. J., Monroe, R. E., Sweeley,
C.G., and Bieber, L.L. (1970) Characterisation of
invertebrate sphingolipid bases: occurrence of
eicosasphing-4,11-diene and eicosasphing-11-enine in
scorpion. Biochim. biophys. Acta 202: 195-197.

Pattnaik, N.M., Mundall, E.C., Trambusti, G.C., Law, J.H.
and Kezdy, F.J. (1979) Isolation and characterization of a larval lipoprotein from haemolymph of Manduca sexta. Comp. Biochem. Physiol. 63B: 469-476.

Peled, Y. and Tietz, A. (1975) Isolation and properties of a lipoprotein from the haemolymph of the locust, Locusta migratoria. Insect Biochem. 5: 61-72.

Ramabrahman, P. and Subrahmanyah, D. (1979) Choline kinase of Culex pipiens fatigans. Insect Biochem. 9: 315-322.

Rao, R.P. and Agarwal, H.C. (1971) Lipids of Trogoderma. III. Phospholipid and fatty acid composition during development with note on fatty acid synthesis. Comp. Biochem. Biophys. 39B: 183-194.

Rao, R.H. and Subrahmanyam, D. (1969) Studies on the phospholipase A in larvae of Culex pipiens fatigans. J. Insect Physiol. 15: 149-159.

Robb, R., Hammond, R. and Bieber, L.L. (1972) Temperature dependent changes in sphingosine composition and composition of fatty acids of glycerophosphatides from Musca domestica larvae. Insect Biochem. 2: 131-136.

Shelley, R.M. and Hodgson, E. (1970) Biosynthesis of phosphatidylcholine in the fat body of Phormia regina larvae. J. Insect Physiol. 16: 131-139.

Shelley, R.M. and Hodgson, E. (1971a) Choline kinase from the fat body of Phormia regina larvae. J. Insect Physiol. 17: 545-558.

Shelley, R.H. and Hodgson, E. (1971b) Substrate specificity and inhibition of choline and ehtanolamine kinases from the fat body of Phormia regina larvae. Insect Biochem. 1: 149-156.

Sohal, R.S. and Bridges, R.G. (1977) Effects of experimental alterations in phospholipid composition on the size and numbers of mitochondria in the flight muscles of the housefly, Musca domestica. J. Cell Sci. 27: 273-287.

Strickland, K.P. (1973) The chemistry of phospholipids, from "Form and Function of Phospholipids" (eds. Ansell, G.B., Hawthorne, J.N. and Dawson, R.M.C.) 1973. Elsevier Scientific Publishing Company Amsterdam-London-New York. 9-42.

Strunecka, A., Markos, A. and Kubista, V. (1978) Differential effect of anaerobiosis on phospholipid metabolism in insect muscle. Insect Biochem. 8: 189-191.

Taylor, J. and Hodgson, E. (1965) The origin of phospholipid ethanolamine in the blowfly, Phormia regina (Meig.). J. Insect Physiol. 11: 281-285.

Thomas, K.K. and Gilbert, L.I. (1968) Isolation and characterisation of the haemolymph lipoproteins of the American silk moth, Hyalophora cecropia. Arch. Biochem. Biophys. 27: 512-521.

Thompson, A.C., Davis, F.M., Henson, R.D., Gueldner, R.C., Hedin, P.A. and Henderson, C.A. (1973a) Lipids

and fatty acids of the south western corn borer
Diatraea grandiosella. J. Insect Physiol. 19:
1817-1823.

Thompson, A.C., Henson, R.D., Gueldner, R.S. and Hedin,
P.A. (1973b) Constituents of the boll weevil, Anthonomus grandis Bohemia VIII. Lipid and fatty acids
in the subcellular particles of pharate pupae. Comp.
Biochem. Physiol. 45B: 233-239.

Wang, C.M. and Patton, R.L. (1969) Lipids in the haemolymph of the cricket, Acheta domesticus. J. Insect
Physiol. 15: 851-860.

White, D.A. (1973) The phospholipid composition of
mammalian tissues from "Form and Function of phospholipid" (eds. Ansell, G.B., Hawthorne, J.N. and Dawson, R.M.C.) Elsevier Scientific Publishing Company.
Amsterdam-London-New York. 441-482.

Willis, N.P. and Hodgson, E. (1970a) Phospholipids and
their constituent fatty acids in two populations of
Dendroctonus frontalis (Coleoptera: Scolytidea).
Ann. Ent. Soc. Amer. 63: 1585-1591.

Willis, N.P. and Hodgson, E. (1970b) Absence of trans
methylation reactions involving choline, betaine and
methionine in the insects. Int. J. Biochem. 1: 659-
662.

Wirtz, K.W.A. (1974) Transfer of phospholipids between
membranes. Biochim. biophys. Acta. 344: 95-117.

Wlodawer, P. and Wisniewska, A. (1965) Lipids in the
haemolymph of wax moth larvae during starvation. J.
Insect Physiol. 11: 11-20.

Yadava, R.P.S. and Musgrave, A.J. (1972) Phospholipid
patterns of two symbiote-harbouring weevils, the rice
weevil Sitophilus oryzae L. and corn weevil Sitophilus zeamais (Mots.) (Coleoptera: Curculionidae).
Comp. Biochem. Physiol. 42B: 197-200.

Yadava, R.P.S., Rattray, J.B.M. and Musgrave, A.J. (1972)
Fatty acid profiles of two microbiologically differ-
ent strains of granary weevil: Sitophilus granarius
L. Coleoptera. Comp. Biochem. Physiol. 43B: 383-391.

Yanagawa, H.-A. and Horie, Y. (1978) Activating enzyme
of phosphorylase B in the fat body of the silkworm
Bombyx mori. Insect Biochem. 8: 155-158.

Yurkiewicz, W. (1967) Phospholipid metabolism in the
embryo of the Indian meal moth, Plodia interpunctella. Proc. Pa. Acad. Sci. 41: 27-29.

Zielinska, Z.M. and Dominas, H. (1967) The origin of
phospholipid ethanolamine and choline in a sawfly
Acantholyda nemoralis. J. Insect Physiol. 13: 1769-
1779.

10
Lipid Transport in Insects

*D. J. Van der Horst**

INTRODUCTION

Lipids comprise significant components in insect nutrition. Definite requirements for lipids in the diet of insects have been stressed by several authors, indicating that--among other constituents--polyunsaturated fatty acids and sterols are essential (for reviews see Gilbert, 1967; House, 1974; Dadd, 1977; Downer, 1978). In addition, dietary lipids, particularly triacylglycerols, provide an important source of metabolic energy.

Upon digestion and absorption in the midgut, exogenous lipids are transported in the haemolymph to sites for utilization or storage. In this respect, the insect fat body occupies a central position in acting both as a metabolic center and a storage organ for lipids and other substrates. So, the lipids encountered in the haemolymph shortly after feeding may predominantly originate from the gut wall, whereas in postabsorptive stages, the transported lipids are derived principally from the fat body lipid stores. Furthermore, in insects which rely on lipids as the major fuel for locomotion, particularly flight, a specialized and intense transport of lipids from the fat body to the flight muscles comes into prominence during flight activity, whereas in female insects which utilize lipids for egg production, a transfer of lipids from the fat body to the ovaries may be the dominant lipid-transporting process in the haemolymph during ovarial development. Of particular interest, therefore, is the continuing presence of diacylglycerols as the dominant neutral lipid class in the haemolymph of all but a few species of insects (for reviews see Gilbert and Chino, 1974; Gilbert et al., 1977; Downer, 1978; Beenak-

---

* Laboratory of Chemical Animal Physiology, State University of Utrecht, 3508 TB Utrecht, The Netherlands.

kers et al., 1981a), as this introduces a fundamental
deviation from the mammalian lipid transport system.

In mammalian plasma, circulating lipids destined for
uptake and storage by adipocytes are transported as chylomicrons or very low density lipoproteins (VLDL). Dietary triacylglycerols, which are hydrolized to 2-mono-
acylglycerols and free fatty acids and after transport
through the enterocyte brush border membrane are resynthesized to triacylglycerols in the mucosal cells of the
intestine (Johnston, 1970), are carried by chylomicrons,
which function specifically to transport dietary lipids
to non-hepatic tissues. In addition, VLDL contain triacylglycerols which are synthesized in the liver (for a
review see Smith et al., 1978). On the other hand,
lipids leaving adipose tissue, for instance during prolonged muscular exercise, are in the form of non-esteri-
fied fatty acids covalently bound to albumin.

In insects, the products of triacylglycerol digestion
appear not to be significantly resynthesized to triacylglycerols in the gut, nor are chylomicrons formed and
released into the haemolymph. Instead, in the few species studied, diacylglycerols are released from the gut
(Weintraub and Tietz, 1973, 1978; Turunen, 1975; Hoffman
and Downer, 1976, 1979a; Chino and Downer, 1979) and
transported bound to specific lipoproteins. Likewise,
though lipids in the fat body are accumulated as triacylglycerols, lipid release from the fat body in many insect
species is in the form of diacylglycerols which are also
taken up by these lipoproteins for transport in the haemolymph (for reviews see Gilbert and Chino, 1974; Gilbert
et al., 1977; Beenakkers et al., 1981a). Finally, even
in a condition of strongly accelerated lipid mobilization
for flight muscle energy supply during insect flight
(which will be discussed below), lipoprotein-bound diacylglycerols remain the principal mode of lipid transport in the haemolymph.

In the present account it is proposed to summarize a
few significant early findings in order to discuss recent
contributions to our knowledge on lipid transport in insects. As diacylglycerol transport particularly will be
detailed, it seems appropriate to state at the outset
that although this may be the major means of lipid transport in the Lepidoptera, Orthoptera, Hemiptera, and Dic-
tyoptera that have been studied, it would be premature
to extrapolate the results to all families and species of
insects. For instance, the waxmoth, Galleria mellonella,
which feeds on the wax of bee hives, has a very specialized digestion and absorption of this substrate as discussed by Wlodawer et al. (1966) and Dadd (1970). Emphasis will be placed on lipid transport during flight, in
which particularly the locust has been studied extensively and will be adopted as a model system. Lastly, since

the deviation of the insect lipid transport system from the mammalian situation suggests important differences in structural features of insect and mammalian lipoproteins, a comparison between both systems will be made.

General aspects of storage, metabolism, and transport of lipid reserves in insects have recently been covered by the reviews of Gilbert and Chino (1974), Bailey (1975), Downer and Matthews (1976), Gilbert et al. (1977), Downer (1978), and Beenakkers et al. (1981a).

TRANSPORT FORMS OF DIACYLGLYCEROL

Diacylglycerol release from insect fat body as a possible means of lipid transport in insects was first established by Chino and Gilbert (1964) in the silkmoth, Hyalophora cecropia. Interestingly, not only saturniid moths which do not feed as adults utilize lipids as their fuel for flight (Domroese and Gilbert, 1964), but also in sugar-feeding adult Lepidoptera diacylglycerol is the principal means of lipid transport (Bhakthan and Gilbert, 1970). With respect to the nature of the haemolymph diacylglycerols it appears that 1,2-diacylglycerols are released from the fat body of the locust (Tietz, 1967; Tietz et al., 1975). It is of particular interest that the 1,2-diacylglycerols isolated from the haemolymph of resting adult Locusta migratoria are stereospecific, revealing the sn-1,2-configuration with a remarkably high optical purity (Tietz and Weintraub, 1980; Lok and Van der Horst, 1980)(Fig. 1). Besides, the elevated haemo-

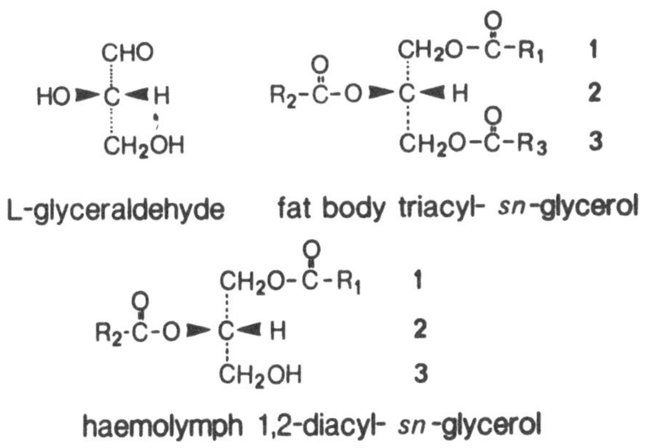

L-glyceraldehyde   fat body triacyl- sn-glycerol

haemolymph 1,2-diacyl- sn-glycerol

FIGURE 1. Stereospecific configuration of locust haemolymph 1,2-diacylglycerols.

lymph 1,2-diacylglycerol level resulting from triacylglycerol mobilization from the fat body induced by flight
or injection of the adipokinetic hormone (which will be
discussed later on) appeared to comprise over 97% of the
sn-1,2-enantiomer in both these cases (Lok and Van der
Horst, 1980). These data indicate stereospecificity of
the processes involved in the production of the 1,2-di-
acylglycerols in the fat body and other tissues expected
to contribute to the haemolymph diacylglycerols, such as
the gut. Thus, either a stereospecific lipase acting on
the sn-3 position of the triacylglycerols might be present, or after degradation of the triacylglycerols to 2-
monoacylglycerols--as in mammalian intestine--diacylgly-
cerols might be synthesized by a stereospecific monoacylglycerol acyltransferase. Evidence for either pathway
is inconclusive (Lok and Van der Horst, 1980); the in
vitro studies by Tietz et al. (1975) and Tietz and Weintraub (1980) on fat body microsomal preparations, however, would favor the latter pathway to be operative in
vivo.

Interestingly, end product specificity of triacylglycerol lipases in the American cockroach, Periplaneta
americana, which unlike the Lepidoptera and the locust
does not depend on a supply of lipids during flight activity (Downer and Matthews, 1977), indicates that in the
midgut 2-monoacylglycerols and free fatty acids are produced for absorption across the intestinal wall, where-
after monoacylglycerol acyltransferase activity in the
intestinal mucosa (Hoffman and Downer, 1979a) results in
the production of a racemic mixture of sn-1,2- and sn2,
3-diacylglycerols (Hoffman and Downer, 1979b) which is
taken up by the haemolymph (Chino and Downer, 1979).
Incubation of cockroach fat body homogenate with triacylglycerol as a substrate primarily yielded 1,2-diacylgly-
cerols with only a slight preference for the sn-1,2-
enantiomer (less than 60%) over the sn-2,3-enantiomer
(Hoffman and Downer, 1979b). Although the exclusive
occurrence of the sn-1,2-enantiomer in the haemolymph of
the locust remains to be explained, it may tentatively
be related to the utilization of diacylglycerol during
flight activity; the stereospecific component may be
required for activation of the flight muscle lipase.

HAEMOLYMPH DIACYLGLYCEROL-TRANSPORTING LIPOPROTEINS

Both the diacylglycerol released from triacylglycerol digestion in the alimentary tract and the diacylglycerol released from the fat body must be transported
in the aqueous haemolymph and therefore are not present
as free molecules, but rather as components of macromole-
cular lipoprotein complexes. These haemolymph lipoproteins have important functions in both release and trans-

port of diacylglycerols. In vitro studies on the release
of diacylglycerols from insect fat body have shown that
the presence of haemolymph lipoproteins is a requisite
for acylglycerol release (Tietz, 1962, 1967; Chino and
Gilbert, 1964, 1965). Haemolymph lipoproteins have been
identified in several species of insects, particularly
in Lepidoptera (e.g. silkmoths), locusts, and cockroaches
(for a review see Wyatt and Pan, 1978).

In the Lepidoptera, a system of two major diacylglycerol-transporting haemolymph lipoprotein fractions is
apparent which--apart from its intrinsic interest--may
be considered as a model system for a number of other
insects. By ammonium sulphate precipitation and DEAE
cellulose column chromatography, Chino et al. (1967,
1969) were the first to characterize the lipoproteins in
the haemolymph of the silkmoth, Philosamia cynthia, which
on the basis of function were termed diacylglycerol-
carrying lipoproteins I and II (DGLP-I and DGLP-II). For
the equivalent lipoproteins isolated by ultracentrifuga-
tion from two related species of silkmoths, H. cecropia
and H. gloveri, Thomas and Gilbert (1968, 1969) on the
basis of density used the terms high density lipoproteins
(HDL) and very high density lipoproteins (VHDL), respectively, by analogy to mammalian serum lipoproteins. For
reasons of clarity, DGLP-I (HDL) and DGLP-II (VHDL) were
renamed LP-I and LP-II (Gilbert and Chino, 1974).

LP-I functions in taking up diacylglycerols from the
fat body and carrying the lipid to other tissues, including the flight muscles and the ovary (Chino et al., 1969,
1977). LP-II appeared to be the female-specific vitellogenin (Chino, 1976) which is selectively sequestered by
the maturing oocytes. This lipoprotein fraction also
carries diacylglycerols, but rather than accepting them
directly from the fat body it accepts diacylglycerols
from LP-I. LP-I thus appears to play a vital role in
lipid transport, including fat mobilization during flight
of the adult insect, whereas LP-II provides substrate to
the developing egg, where it is used as both an energy
source and a structural material during subsequent
embryogenesis. However, as stressed by Chino et al.
(1977), the major source of lipid for vitellogenesis is
provided by LP-I.

Mol. wt of the globular LP-I is about 700,000 daltons
and of LP-II approx. 500,000 daltons (Chino et al., 1969;
Pan and Wallace, 1974). Lipid content of LP-I is 44-48%
and of LP-II 6-10%, diacylglycerols comprising almost 50%
of both lipoprotein species (Gilbert et al., 1977).
Though very small amounts of triacylglycerols, sterol
esters, and free fatty acids are present, substantial
amounts of phospholipids and free sterol indicate that
apart from diacylglycerol transport, both lipoprotein
fractions may be implicated in a transport of phospho-

lipids and sterol. Thus, for example, LP-I can elicit cholesterol release from isolated midgut preparations in vitro, while it appeared that cholesterol--like diacyl-glycerol--could be transferred from LP-I to LP-II (Chino and Gilbert, 1971). It should be noted that although di-acylglycerol release from the fat body is clearly impor-tant to energy metabolism, sterol release from the gut is of even greater importance as insects lack sterol syn-thetic capacity and require dietary sterols for a multi-plicity of vital processes, including synthesis of the molting and vitellogenic hormone, ecdysone (see Gilbert, 1967; Gilbert et al., 1977; Downer, 1978).

In addition to this system of two high density lipo-protein fractions, a third, low density lipoprotein (LDL) has been isolated from H. cecropia (Thomas and Gilbert, 1968), suggesting that an integrated view of lipid trans-port must be more complex than outlined above.

In general, the lepidopteran system of diacylglycerol -carrying lipoproteins also accommodates the fragmentary information from the few other insect species investigat-ed. In the cockroach, Leucophaea maderae, a vitellogenin has been isolated and purified (Dejmal and Brookes, 1972; Engelmann et al., 1976). In addition, in P. americana, another DGLP (mol. wt approx. 600,000) from the haemo-lymph of adult male and female insects has been reported recently (Chino et al., 1981), and appears to be analo-gous to LP-I from the silkmoth. Interestingly, it was demonstrated that the same lipoprotein served to trans-port the diacylglycerol both from the site of storage (i.e. the fat body) and from the site of absorption (e.g. the gut), which also suggests that the same mechanism of diacylglycerol release and uptake is involved in the two processes in this group of insects. This is in obvious disagreement with earlier observations from Reisser-Bollade (1976; see also Hoffman and Downer, 1979a). As cockroach DGLP, in addition to diacylglycerols and cho-lesterol, contained even higher amounts of hydrocarbons, the composition of which was essentially similar to the cuticular hydrocarbons, multiple lipid-transporting func-tions of the lipoprotein in the cockroach were suggested (Chino et al., 1981).

Even in the larval waxmoth, which differs consider-ably from other insects in lipid mobilization and trans-port as indicated above, Thomas (1979) identified two high density lipoproteins (LP-Ia and LP-Ib) and a very high density lipoprotein (LP-II), which in general resem-bled the lipoproteins of other Lepidoptera in lipid com-position and contents as well as in other properties. About 70% of the lipid in the haemolymph is bound to LP-Ia and in all three lipoproteins diacylglycerol is the predominant lipid (approx. 50%), but appreciable amounts of free fatty acids (9-14%) are also present. Unfortunately, however, in this study the larvae were reared on an artificial diet relatively poor in lipid compared to beeswax.

In the migratory locust, a female-specific diacylglycerol-carrying vitellogenin (LP-II) in the haemolymph has been purified and characterized (Gellissen et al., 1976; Chen et al., 1976, 1978; Harry et al., 1979; Chinzei et al., 1981). In addition, diacylglycerol is carried by sex-unspecific lipoprotein fractions (Peled and Tietz, 1975; Gellissen and Emmerich, 1978). So, it would appear tempting to extend the analogy of the lepidopteran lipid system to locusts, which in fact has been proposed by Gellissen and Emmerich (1980). This would, however, seem premature as in the locust remarkable changes in the haemolymph lipoprotein pattern occur during increased lipid mobilization induced by flight activity or injection of the adipokinetic hormone (as will be detailed below), which have not (yet?) been observed in other insect species, though in migratory Lepidoptera an elevation of the haemolymph lipid level during flight as well as the presence of an adipokinetic hormone have been reported recently (Dallmann and Herman, 1978; Herman and Dallmann, 1981; Turunen and Chippendale, 1981). Moreover, the interchange of lipids between lepidopteran LP-I and LP-II has not been observed between the locust LP-I-like diacylglycerol-carrying lipoproteins and the female-specific vitellogenin (LP-II), though in Locusta both vitellogenin and DGLP apparently are incorporated into oocytes (Lubzens et al. 1981).

Hence, for the present, the lipid transport in the locust may be considered to differ in some respects from that in other, mainly lepidopteran, insect species. In particular, the lipoprotein-bound diacylglycerol transport during flight, which has been studied only in locusts and provides some important information on the mechanism of flight-induced lipid transport, is discussed here in more detail.

TRANSPORT OF DIACYLGLYCEROLS DURING FLIGHT

In the migratory locust it has long been recognized that the main source of energy for flight muscle contraction during sustained flight is diacylglycerol, which is mobilized from fat body triacylglycerol stores and transported in the haemolymph (for reviews see Beenakkers et al., 1981a,b). During prolonged flight, the level of haemolymph diacylglycerols is elevated about threefold, while turnover rate is accelerated almost 9-fold (Van der Horst et al., 1978a). Following the initiation of flight, the relative contribution of fatty acid oxidation to the energy-generating processes in the flight muscles increases progressively (Van der Horst et al., 1980);

190

after some 2 h of flight, blood-borne diacylglycerols
account for nearly 80% of the energy production (Van der
Horst et al., 1978b).

The adipokinetic hormone, synthesized in and released
from the corpus cardiacum, is involved in the process of
flight-induced lipid mobilization. Injection of the hormone into resting locusts evokes a rapid elevation of the
haemolymph diacylglycerol level (for reviews see Mordue
and Stone, 1979; Beenakkers et al., 1981a,b).

The specific haemolymph lipoproteins required for the
the release of diacylglycerols from the locust fat body
(Tietz, 1962, 1967) are synthesized in the fat body; inhibition of protein synthesis in vitro, however, did not
affect the release of diacylglycerols nor the uptake of
lipid by the lipoprotein, which indicates that the haemolymph lipoproteins can be loaded independently of de novo
protein synthesis (Peled and Tietz, 1973).

During increased lipid mobilization from the fat body
for flight muscle energy supply, not only the haemolymph
lipid profile is altered, but the pattern of haemolymph
lipoproteins carrying the elevated diacylglycerol is
changed as well (Mayer and Candy, 1967; Van der Horst et
al., 1979; Mwangi and Goldsworthy, 1981). By gel filtration chromatography of haemolymph of resting adult male
L. migratoria, Mwangi and Goldsworthy (1977) eluted a
yellow high mol. wt lipoprotein fraction (A) to which
diacylglycerol was bound, accompanied by two non-lipid
containing protein fractions (B and C). Injection of

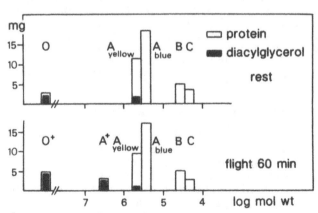

FIGURE 2. Schematic representation of the haemolymph
(lipo)protein fractions of the locust, isolated by gel
filtration of 1.0 ml haemolymph samples on Ultrogel
AcA22. Fractions 0 and 0$^+$ were eluted at the void
volume of the column.

adipokinetic hormone resulted in the appearance of a new lipoprotein fraction of higher mol. wt (A^+), to which the majority of the elevated diacylglycerol was bound, whereas fraction A decreased. More recently, in addition to lipoprotein fraction A, which turned out to be composed of a diacyl-glycerol-carrying yellow lipoprotein (A_{yellow}) and a blue protein fraction containing no lipid (A_{blue}), a second lipoprotein fraction (0) of higher mol. wt appeared to carry diacylglycerols (Van der Horst et al., 1979) (Fig. 2). Lipoprotein fraction 0 was not retained by the gel column; mol. wt of lipoprotein A_{yellow} is approx. 45,000. Elevation of the haemolymph diacylglycerol level by flight or injection of the adipokinetic hormone resulted apart from formation of A^+ (mol. wt about 3,500,000) in an increase in both lipid and protein contents of fraction 0 (Fig. 2). These changes in the haemolymph lipoproteins during flight, resulting in a higher capacity for diacylglycerol uptake and transport in the haemolymph, proved to be essential for the progressive turnover rate of diacylglycerol during flight (Beenakkers et al., 1978; Van der Horst et al., 1979). Interestingly, unlike lepidopteran LP-I which is a high density lipoprotein (as mentioned above), fraction A^+ appeared to be composed of low density lipoprotein (LDL) particles resembling human LDL lipoproteins (Goldsworthy and Wheeler, 1981).

Total protein concentration in the haemolymph remained unaffected during flight, so it seemed reasonable to assume that the changes in the lipoprotein pattern originate completely from associations of protein fractions present in the haemolymph at rest.

Immunoelectrophoresis and immunodiffusion employing monospecific antisera prepared against the isolated and purified fractions 0 and A_{yellow} indicated an immunological identity between both lipoprotein fractions; the very high mol. wt fraction 0 apparently is a polymer of possibly modified lipoprotein A_{yellow} (Van der Horst et al., 1981a). In the lipoprotein fractions A^+ and 0^+ appearing upon injection of locusts with adipokinetic hormone, three protein bands were separated electrophoretically, one of them containing lipids. Two protein bands are immunologically identical with basic lipoprotein A_{yellow}, whereas the third protein component of both lipoprotein fractions is identical with the nonlipid carrying haemolymph protein fraction C (Van der Horst et al., 1981a).

These data strongly supported the possibility of reassociations of haemolymph protein fractions evoked by the adipokinetic hormone. Actual dynamics in these conversions were studied with radioiodinated lipoprotein A_{yellow} (of which specifically the apoprotein was labelled) and iodinated protein fraction C.

Injection of $[^{125}I]$-A_{yellow} into resting locusts resulted in an immediate exchange of apoprotein between A_{yellow} and 0. Moreover, simultaneous injection of A_{yellow} and synthetic adipokinetic hormone led to a heavy labelling of both lipoprotein A^+ and 0^+, indicative of a transfer of apolipoprotein A_{yellow} to both elevated lipoproteins (Van der Horst et al., 1981b). From experiments with radioiodinated protein C it was likewise inferred that protein C participates in the formation of both A^+ and 0^+. The increased amount of diacylglycerols carried by the latter fractions is principally derived from the elevated release of diacylglycerols from the fat body. However, upon specific labelling of the diacylglycerols bound to A_{yellow} and 0 with $[^{14}C]$-fatty acids, a transfer of labelled diacylglycerols from A_{yellow} to both higher mol. wt lipoproteins was evidenced, indicative of participation of the lipid component from A_{yellow} in the lipoprotein reorganizations as well (Van der Horst et al., 1981b).

Thus, the locust provides a unique concept of lipoprotein remodelling, in which haemolymph protein and lipoprotein components already existing in the resting stage give rise to new lipoprotein species capable of accepting and carrying the increased amount of diacylglycerols mobilized by the fat body in response to flight activity.

Though the net consequence of the dynamic mutual interactions between released lipids and circulating haemolymph proteins and lipoproteins in the locust is that lipids are efficiently delivered to the contracting flight muscles, at present some questions remain unanswered. For instance, it is not known why protein fraction C participates in the lipoprotein fractions elevated during flight. One explanation might be that participation of protein C enables the remarkable increase in the lipid-loading capacity of the lipoproteins. At rest, the diacylglycerol:protein ratio of lipoprotein A_{yellow} and fraction 0 are 0.18 and 0.78, respectively. After flight, the ratio for A_{yellow} is slightly reduced, but the diacylglycerol:protein ratio of A^+ is 1.36 and that of 0^+ is increased to 1.10 (Van der Horst et al., 1979; Beenakkers et al., 1981a). Another explanation, which has been advanced by Wheeler and Goldsworthy (1981) may be the possibility of specific interactions of protein C with receptors in the flight muscles. The latter suggestion would include a possible activation of the flight muscle (lipoprotein) lipase by protein C resembling the activation of mammalian lipoprotein lipase by (purely coincidental!) apoprotein C (see Smith et al., 1978).

Furthermore, from a teleological point of view, it is not clear why in the (male) locust two lipoprotein fractions carry diacylglycerol at rest and three during

flight. Very recent experimental results on the conver-
sions of $|^{125}I|$-labelled lipoproteins during flight
give evidence that particularly the diacylglycerols carried by lipoprotein A^+ are utilized for flight muscle
energy supply (resulting in recovery of the label in both
A_{yellow} and C) (Van der Horst, Storm, Van Doorn and
Beenakkers, to be published), which confirms the suggestion by Mwangi and Goldsworthy (1981). However, the
function of the slightly opalescent chylomicron-like
fraction 0, carrying about half of the total amount of
haemolymph diacylglycerols in the resting stage, and--
even more intriguing--that of 0^+ (including the participation of protein fraction C) remain as yet unexplained. Since recovery of fraction 0 (and 0^+) after
gel filtration chromatography of haemolymph samples is
largely prevented by treatment of haemolymph with ammonium sulphate at a final concentration of 50% (Van der
Horst et al., 1979), it would be easiest to suggest that
this fraction represents denatured (lipo)protein(s)
(Mwangi and Goldsworthy, 1981). However, since the pre-
cipitated protein residue resulting from haemolymph
treatment with ammonium sulphate appeared to contain
significant amounts of diacylglycerols (Van der Horst et
al., 1979), and additionally, in view of both biochemical
parameters such as the lipid : protein ratio and the obvious dynamic behavior of fraction 0 as described above,
such a suggestion would seem an oversimplification.

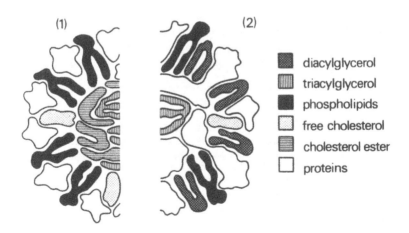

FIGURE 3. Structural models of human (1) (after Edelstein et al., 1979) and insect (2) lipoproteins.

STRUCTURAL FEATURES OF INSECT LIPOPROTEINS

Comparison of structural features of mammalian and insect lipoproteins reveals important differences and suggests that the lipoprotein particles must be constructed according to different principles. In the mammalian system, triacylglycerols and sterol esters tend to occupy the apolar core of the lipoprotein particle, which is stabilized by a monolayer composed of phospholipids, cholesterol, and apoprotein (Scanu, 1978). From the correlations between size and composition of human serum lipoproteins, a unifying concept of a general structural model has been inferred (Shen et al., 1977; Edelstein et al., 1979). According to the proposed model, both apoprotein polypeptide chains and hydrophylic head groups of phospholipids are closely packed at the outer surface of the particle, whereas the polar head groups of free cholesterol are located deeper in the monolayer and occupy an area beneath the apolipoproteins (Fig. 3).

In the insect system, the polar nature of the diacylglycerols would relegate them to the surface of the lipoprotein particle (Pattnaik et al., 1979; Law, 1980), which may account for the high rate of diacylglycerol exchange among the different lipoprotein species in insect haemolymph (Van der Horst et al., 1981b), being in contrast to the poor exchange of triacylglycerols between mammalian lipoproteins (Bell, 1978). Phospholipids and cholesterol may also be at or near the surface of the insect lipoprotein. Non-polar lipids, especially triacylglycerols and sterol esters, are greatly reduced or virtually absent (Gilbert et al., 1977; Downer and Chino, 1979). From calculations based upon particle size and lipid and apoprotein composition of a haemolymph lipoprotein isolated from larvae of the tobacco hornworm, Manduca sexta, Pattnaik et al. (1979) concluded that the central core region of the spherical lipoprotein is far too large for the small amount of non-polar lipids present and suggested that a significant portion of the apoprotein may lie inside the particle, which would harmonize with findings from degradation of the lipoprotein by proteolytic enzymes and solubility characteristics. A model for such an insect lipoprotein, consistent with the space-filling requirements and the proposed localization of the known components, is suggested in Fig. 3. In view of the lack of details, this model is largely speculative.

Interestingly, amino acid analyses performed on the total apoproteins of the haemolymph diacylglycerol-carrying lipoproteins isolated from different insect species show a remarkable conformity in apoprotein amino acid compositions (Table 1). The preliminary data on the locust lipoproteins (Van der Horst, 1981) suggest that

TABLE 1. Amino Acid Composition of total apoprotein of diacylglycerol-carrying lipoproteins from some insect species

	Silkworm[a] LP-1	Cockroach[b]	Mole % Tobacco hornworm[c] (larvae)	Locust[d] 0	Ayellow	0+	A+
Asp	12.6	11.0	12.5	12.3	12.2	12.1	12.2
Thr	4.9	6.6	3.9	5.8	5.6	6.1	6.2
Ser	6.9	6.9	7.7	7.2	6.9	6.5	6.1
Glu	10.4	10.8	9.7	12.8	11.9	15.8	15.4
Pro	4.7	3.8	5.0	5.1	4.9	3.9	3.9
Gly	6.7	6.4	6.6	6.7	6.2	4.9	4.5
Ala	6.3	6.8	7.5	7.0	7.4	10.8	12.1
Val	7.4	8.4	7.1	7.1	8.0	6.6	6.5
1/2 Cys		0.6		0.1	0.5	0.5	0.5
Ile	5.8	4.1	5.4	5.2	5.6	5.0	4.9
Leu	9.0	10.7	8.7	10.4	10.9	11.1	11.0
Tyr	2.8	3.0	3.8	0.1	1.3	0.8	1.4
Phe	4.8	4.7	5.2	4.0	5.0	3.2	3.2
Lys	10.7	9.3	8.5	8.1	8.1	7.3	7.4
His	2.8	3.9	2.9	5.4	2.2	3.0	3.0
Arg	3.7	2.7	4.1	2.9	3.4	2.4	2.2
Met	0.5	0.3	1.3				

aFrom Gilbert and Chino (1974). bFrom Chino et al. (1981). cFrom Pattnaik et al. (1979). dFrom Van der Horst (1981); amino acids that change in response to adipokinetic hormone are underlined (see text).

injection of adipokinetic hormone evokes an increase in glutamate and alanine, concomitant with some decrease in other amino acids such as proline, glycine, phenylalanine, lysine, and arginine in the elevated lipoprotein fractions A^+ and 0^+, which may be accounted for by participation of protein fraction C. Indeed, protein C appeared to contain very high levels of both glutamate (about 20%) and alanine (about 16%) (Van der Horst and Van Doorn, unpublished results).

In summary, essential structural differences between insect and mammalian lipoproteins are the substitution for a substantial proportion of the neutral lipids in the apolar core by the apoproteins and the localization of diacylglycerols in the easily accessible shell. Particularly the presence of the diacylglycerols at the surface of the insect lipoprotein particle is a unique feature, which obviously allows for the rapid uptake of diacylglycerols from the fat body as well as the high rate of transfer of diacylglycerols from the circulating lipoproteins to the flight muscles without degradation of the rest of the particle, thus enabling the lipoproteins to operate as a shuttle mechanism which is of vital importance during increased lipid mobilization for flight muscle energy supply, as discussed by Gilbert and Chino (1974) and Van der Horst (1981).

CONCLUSION

Though the present communication is inevitably rather selective in approach, it has been possible to discuss our present knowledge of phenomena implicated in the transport of some major lipid components (particularly diacylglycerols) in the haemolymph of insects, revealing fundamental differences from lipid transport as postulated for mammals.

REFERENCES

Bailey, E. (1975) Biochemistry of insect flight. 2. Fuel supply. In Insect Biochemistry and Function (Ed. by Candy, D.J. and B.A. Kilby). pp. 89-176. Chapman and Hall, London.

Beenakkers, A.M.Th., D.J. Van der Horst, and W.J.A. Van Marrewijk. (1978) Regulation of release and metabolic function of the adipokinetic hormone in insects. In Comparative Endocrinology (Ed. by P.J. Gaillard and H.H. Boer). pp. 445-448. Elsevier/North-Holland Biomedical Press, Amsterdam.

Beenakkers, A.M.Th., D.J. Van der Horst and W.J.A. Van Marrewijk. (1981a) Role of lipids in energy metabolism. In Energy Metabolism and its Regulation in Insects (Ed. by R.G.H. Downer). pp. 53-100. Plenum

Press, New York.

Beenakkers, A.M.Th., D.J. Van der Horst, and W.J.A. Van Marrewijk. (1981b) Metabolism during locust flight. Comp. Biochem. Physiol. 69B: 315-321.

Bell, F.P. (1978) Lipid exchange and transfer between biological lipid-protein structures. Prog. Lipid Res. 17: 207-243.

Bhakthan, N.M.G. and L.I. Gilbert. (1970) Studies on lipid transport in Manduca sexta (Insecta). Comp. Biochem. Physiol. 33: 705-706.

Chen, T. T., P. Couble, F. L. De Lucca and G. R. Wyatt. (1976) Juvenile hormone control of vitellogenin synthesis in Locusta migratoria. In The Juvenile Hormones (Ed. by L.I. Gilbert). pp. 505-529. Plenum Press, New York.

Chen, T. T., P. W. Strahlendorf and G. R. Wyatt. (1978) Vitellin and vitellogenin from locusts (Locusta migratoria). Properties and post-translational modification in the fat body. J. Biol. Chem. 253: 5325-5331.

Chino, H. and R.G.H. Downer. (1979) The role of diacylglycerol in absorption of dietary glyceride in the American cockroach, Periplaneta americana L. Insect Biochem. 9: 379-382.

Chino, H. and L.I. Gilbert. (1964) Diglyceride release from insect fat body: a possible means of lipid transport. Science 143: 359-361.

Chino, H. and L.I. Gilbert. (1965) Lipid release and transport in insects. Biochim. biophys. Acta 98: 94-110.

Chino, H. and L.I. Gilbert. (1971) The uptake and transport of cholesterol by haemolymph lipoproteins. Insect Biochem. 1: 337-347.

Chino, H., A. Sudo and K. Harashima. (1967) Isolation of diglyceride-bound lipoprotein from insect haemolymph. Biochim. biophys. Acta 144: 177-179.

Chino, H., S. Murakami and K. Harashima. (1969) Diglyceride-carrying lipoproteins in insect hemolymph. Isolation, purification and properties. Biochim. biophys. Acta 176: 1-26.

Chino, H., M. Yamagata and K. Takahashi. (1976) Isolation and characterization of insect vitellogenin. Its identity with hemolymph lipoprotein II. Biochim. biophys. Acta 441: 349-353.

Chino, H., R.G.H. Downer and K. Takahashi. (1977) The role of diacylglycerol-carrying lipoprotein I in lipid transport during insect vitellogenesis. Biochim. biophys. Acta 487: 508-516.

Chino, H., H. Katase, R.G.H. Downer and K. Takahashi. (1981) Diacylglycerol-carrying lipoprotein of hemolymph of the American cockroach: purification, characterization, and function. J. Lipid Res. 22: 7-15.

Chinzei, Y., H. Chino and G.R. Wyatt. (1981) Purification and properties of vitellogenin and vitellin from Locusta migratoria. Insect Biochem. 11: 1-7.

Dadd, R.H. (1970) Digestion in insects. In Chemical Zoology (Ed. by M. Florkin and B.T. Scheer). Vol. 5, pp. 117-145. Academic Press, New York.

Dadd, R.H. (1977) Qualitative requirements and utilization of nutrients: insects. In Handbook Series in Nutrition and Food (Ed. by M. Rechcigl). Vol. I, pp. 305-346. CRC Press, Cleveland.

Dallmann, S.H. and W.S. Herman. (1978) Hormonal regulation of hemolymph lipid concentration in the Monarch butterfly, Danaus plexippus. Gen. Comp. Endocrinol. 36: 142-150.

Dejmal, R.K. and V.J. Brookes. (1972) Insect lipovitellin. J. biol. Chem. 247: 869-874.

Domroese, K.A. and L.I. Gilbert. (1964) The role of lipid in adult development and flight-muscle metabolism in Hyalophora cecropia. J. exp. Bio. 41: 573-590.

Downer, R.G.H. (1978) Functional role of lipids in insects. In Biochemistry of Insects (Ed. by M. Rockstein). pp. 57-92. Academic Press, New York.

Downer, R.G.H. and H. Chino. (1979) Cholesterol and cholesterol ester in haemolymph of the American cockroach, Periplaneta americana L. Can. J. Zool. 57: 1333-1336.

Downer, R.G.H. and J.R. Matthews. (1976) Patterns of lipid distribution in insects. Amer. Zool. 16: 733-745.

Downer, R.G.H. and J.R. Matthews. (1977) Production and utilisation of glucose in the American cockroach, Periplaneta americana. J. Insect Physiol. 23: 1429-1435.

Edelstein, C., F.J. Kézdy, A.M. Scanu and B.W. Shen. (1979) Apolipoproteins and the structural organization of plasma lipoproteins: human plasma high density lipoprotein-3. J. Lipid Res. 20: 143-153.

Engelmann, F., T. Friedel and M. Ladduwahetty. (1976) The native vitellogenin of the cockroach Leucophaea maderae. Insect Biochem. 6: 211-220.

Gellissen, G. and H. Emmerich. (1978) Changes in the titer of vitellogenin and of diglyceride carrier lipoprotein in the blood of adult Locusta migratoria. Insect Biochem. 8: 403-412.

Gellissen, G. and H. Emmerich. (1980) Purification and properties of a diglyceride-binding lipoprotein (LP I) of the hemolymph of adult male Locusta migratoria. J. comp. Physiol. 136: 1-9.

Gellissen, G., E. Wajc, E. Cohen, E. Emmerich, S.W. Applebaum and J. Flossdorf. (1976) Purification and properties of oöcyte vitellin from the migratory locust. J. comp. Physiol. 108: 287-301.

Gilbert, L.I. (1967) Lipid metabolism and function in insects. Adv. Insect Physiol. 4: 69-211.

Gilbert, L.I. and H. Chino. (1974) Transport of lipids in insects. J. Lipid Res. 15: 439-456.

Gilbert, L.I., W. Goodman and W.E. Bollenbacher. (1977) Biochemistry of regulatory lipids and sterols in insects. In Biochemistry of Lipids II (Ed. by T.W. Goodwin). Vol. 14, pp. 1-50. University Park Press, Baltimore, Maryland.

Goldsworthy, G.J. and C.H. Wheeler. (1981) Low density lipoproteins in the blood of an insect. Gen. Comp. Endocrinol., in press.

Harry, P., M. Pines and S.W. Applebaum. (1979) Changes in the pattern of secretion of locust female diglyceride-carrying lipoprotein and vitellogenin by the fat body in vitro during oocyte development. Comp. Biochem. Physiol. 63B: 287-293.

Herman, W.S. and S.H. Dallmann. (1981) Endocrine biology of the Painted Lady butterfly Vanessa cardui. J. Insect Physiol. 27: 163-168.

Hoffman, A.G.D. and R.G.H. Downer. (1976) The crop as an organ of glyceride absorption in the American cockroach, Periplaneta americana L. Can. J. Zool. 54: 1165-1171.

Hoffman, A.G.D. and R.G.H. Downer. (1979a) Synthesis of diacylglycerols by monoacylglycerol acyltransferase from crop, midgut and fat body tissues of the American cockroach, Periplaneta americana L. Insect Biochem. 9: 129-134.

Hoffman, A.G.D. and R.G.H. Downer. (1979b) End product specificity of triacylglycerol lipases from intestine, fat body, muscle and haemolymph of the American cockroach, Periplaneta americana L. Lipids 14: 893-899.

House, H.L. (1974) Digestion. In The Physiology of Insecta (Ed. by M. Rockstein). Vol. 5, pp. 63-117. Academic Press, New York.

Johnston, J.M. (1970) Intestinal absorption of fats. In Comprehensive Biochemistry (Ed. by M. Florkin and E.M. Stotz). pp. 1-18. Elsevier, Amsterdam.

Law, J.H. (1980) Lipid-protein interactions in insects. In Insect Biology in the Future (Ed. by M. Locke and D.S. Smith). pp. 295-309. Academic Press, New York.

Lok, C.M. and D.J. Van der Horst. (1980) Chiral 1,2-diacylglycerols in the haemolymph of the locust, Locusta migratoria. Biochim. biophys. Acta 618: 80-87.

Lubzens, E., A. Tietz, M. Pines and S.W. Applebaum. (1981) Lipid accumulation in oöcytes of Locusta migratoria migratorioides. Insect Biochem. 11: 323-329.

Mayer, R.J. and D.J. Candy. (1967) Changes in haemolymph lipoproteins during locust flight. Nature, Lond. 215: 987.

Mordue, W. and J.V. Stone. (1979) Insect Hormones. In Hormones and Evolution (Ed. by E.J.W. Barrington). pp. 215-271. Academic Press, New York.

Mwangi, R.W. and G.J. Goldsworthy. (1977) Diglyceride-transporting lipoproteins in Locusta. J. comp. Physiol. 114: 177-190.

Mwangi, R.W. and G.J. Goldsworthy. (1981) Diacylglyce-rol-transporting lipoproteins and flight in Locusta. J. Insect Physiol. 27: 47-50.

Pan, M.L. and R.A. Wallace. (1974) Cecropia vitello-genin: isolation and characterization. Amer. Zool. 14: 1239-1242.

Pattnaik, N.M., E.C. Mundall, B.G. Trambusti, J.H. Law and F.J. Kézdy. (1979) Isolation and characteriza-tion of a larval lipoprotein from the hemolymph of Manduca sexta. Comp. Biochem. Physiol. 63B: 469-476.

Peled, Y. and A. Tietz. (1973) Fat transport in the locust, Locusta migratoria: the role of protein syn-thesis. Biochim. biophys. Acta 296: 499-509.

Peled, Y. and A. Tietz. (1975) Isolation and properties of a lipoprotein from the haemolymph of the locust, Locusta migratoria. Insect Biochem. 5: 61-72.

Reisser-Bollade, D. (1976) Exogenous lipid transport by the haemolymph lipoproteins in Periplaneta americana. Insect Biochem. 6: 241-246.

Scanu, A.M. (1978) Plasma lipoproteins: structure, function, and regulation. Trends Biochem. Sci. 3: 202-205.

Shen, B.W., A.M. Scanu and F.J. Kézdy. (1977) Structure of human serum lipoproteins inferred from composi-tional analysis. Proc. Natl. Acad. Sci. USA 74: 837-841.

Smith, L.C., H.J. Pownall and A.M. Gotto, Jr. (1978) The plasma lipoproteins: structure and metabolism. A. Rev. Biochem. 47: 751-777.

Thomas, K.K. (1979) Isolation and partial characteriza-tion of the haemolymph lipoproteins of the wax moth, Galleria mellonella. Insect Biochem. 9: 211-219.

Thomas, K.K. and L.I. Gilbert. (1968) Isolation and characterization of the haemolymph lipoproteins of the American silkmoth, Hyalophora cecropia. Archs Biochem. Biophys. 127: 512-521.

Thomas, K.K. and L.I. Gilbert. (1969) The haemolymph lipoproteins of the American silkmoth, Hyalophora gloveri: studies on lipid composition, origin and function. Physiol. Chem. Phys. 1: 293-311.

Tietz, A. (1962) Fat transport in the locust. J. Lipid Res. 3: 421-426.

Tietz, A. (1967) Fat transport in the locust: the role of diglycerides. Eur. J. Biochem. 2: 236-242.

Tietz, A. and H. Weintraub. (1980) The stereospecific structure of haemolymph and fat-body 1,2-diacylglycerol from Locusta migratoria. Insect Biochem. 10: 61-63.

Tietz, A., H. Weintraub and Y. Peled. (1975) Utilization of 2-acyl-sn-glycerol by locust fat body microsomes. Specificity of the acyltransferase system. Biochim. biophys. Acta 388: 165-170.

Turunen, S. (1975) Absorbtion and transport of dietary lipid in Pieris brassicae. J. Insect Physiol. 21: 1521-1529.

Turunen, S. and G.M. Chippendale. (1981) Lipid transport in the migrating Monarch butterfly, Danaus p. plexippus. Experientia 37: 266-268.

Van der Horst, D.J. (1981) Resources and substrate transport. In Exogenous and Endogenous Influences on Metabolic and Neural Control (Ed. by N. Spronk). Pergamon Press, Oxford, in press.

Van der Horst, D.J., A.M.C. Baljet, A.M.Th. Beenakkers and E. Van Handel. (1978a) Turnover of locust haemolymph diglycerides during flight and rest. Insect Biochem. 8: 369-373.

Van der Horst, D.J., J.M. Van Doorn and A.M.Th. Beenakkers. (1978b) Dynamics in the haemolymph trehalose pool during flight of the locust, Locusta migratoria. Insect Biochem. 8: 413-416.

Van der Horst, D.J., J.M. Van Doorn and A.M.Th. Beenakkers. (1979) Effects of the adipokinetic hormone on the release and turnover of haemolymph diglycerides and on the formation of the diglyceride-transporting lipoprotein system during locust flight. Insect Biochem. 9: 627-635.

Van der Horst, D.J., N.M.D. Houben and A.M.Th. Beenakkers. (1980) Dynamics of energy substrates in the haemolymph of Locusta migratoria during flight. J. Insect Physiol. 26: 441-448.

Van der Horst, D.J., P. Stoppie, R. Huybrechts, A. De Loof and A.M.Th. Beenakkers. (1981a) Immunological relationships between the diacylglycerol-transporting lipoproteins in the haemolymph of Locusta. Comp. Biochem. Physiol., in press.

Van der Horst, D.J., J.M. Van Doorn, A.N. De Keijzer and A.M.Th. Beenakkers. (1981b) Interconversions of diacylglycerol-transporting lipoproteins in the haemolymph of Locusta migratoria. Insect Biochem. 11: in press.

Weintraub, H. and A. Teitz. (1973) Triglyceride digestion and absorption in the locust, Locusta migratoria. Biochim. biophys. Acta 306: 31-41.

Weintraub, H. and A. Tietz. (1978) Lipid absorption by isolated intestinal preparations. Insect Biochem. 8: 267-274.

Wheeler, C.H. and G.J. Goldsworthy. (1981) Interaction of haemolymph proteins in Locusta during the action of AKH. Gen. Comp. Endocrinol., In press.

Wlodawer, P., E. Langwinska and J. Baranska. (1966) Esterification of fatty acids in the wax moth haemolymph and its possible role in lipid transport. J. Insect Physiol. 12: 547-560.

Wyatt, G.R. and M.L. Pan. (1978) Insect plasma proteins. A. Rev. Biochem. 47: 779-817.

11
Biosynthesis of Insect Cuticular Hydrocarbons: Application of Carbon-13 NMR Spectroscopy

Gary J. Blomquist
*and Mertxe de Renobales**

INTRODUCTION

Cuticular lipids prevent desiccation in insects, and in some species, serve in chemical communication. The chemistry of insect cuticular lipids has recently been reviewed (Jackson and Blomquist, 1976; Nelson, 1978; Blomquist and Jackson, 1979; and Lockey, 1980). In most species, hydrocarbons are the major lipid class present, and consist of normal, unsaturated, and methyl branched components which range in chain length from 21 to 50+ carbons. The role of cuticular lipids in reducing transpiration has been recently reviewed by Hadley (1980), and the semiochemical functions attributed to hydrocarbons, including those of the cuticular components, have been summarized by Howard and Blomquist (1982). Biosynthetic pathways for the major cuticular lipid components have been established primarily from in vivo studies monitoring the incorporation of radioactive precursors into specific cuticular components (Blomquist and Jackson, 1979; Howard and Blomquist, 1982). Recent studies (Blomquist et al., 1980a; Dwyer et al., 1981b; Howard et al., 1981; Dillwith et al., 1981b) have demonstrated the usefulness of carbon-13 nuclear magnetic resonance (NMR) spectroscopy in studies of the chemistry and biochemistry of insect cuticular lipids. This chapter summarizes information available on the biosynthesis of insect cuticular hydrocarbons, with discussions centered around results from our laboratory on the application of carbon-13 NMR spectroscopy to these studies.

GENERAL CONSIDERATIONS

A number of reviews have described the application of ^{13}C-NMR to biosynthetic studies (McInnes et al.,

---

* Division of Biochemistry, University of Nevada, Reno, Nevada 89557, USA

1976; Kunesch and Poupat, 1977; Grutzner, 1972; Simpson, 1975; Sequin and Scott, 1974; London et al., 1975). A major advantage of ^{13}C-NMR techniques to follow a labeled molecule in biosynthetic work is that the exact location of the labeled carbon can be readily ascertained without the need for degradative studies. This advantage made ^{13}C-NMR attractive for work on the biosynthesis of insect cuticular lipids, since the difficulty in degradative studies of hydrocarbons has discouraged workers from determining the exact location of radioactive atoms. The necessity for reasonably large sample sizes (at least 2 or 3 mg for natural abundance samples and at least 100 µg for significantly enriched samples in the studies on insect cuticular hydrocarbons) and significant enrichment (at least 0.5% for singly labeled products) was met by using a number of insects in which labeled substrate was injected or fed over a several day to several week time period (Dwyer et al., 1981a; 1981b; Blomquist et al., 1981b; Dillwith et al., 1981b).

It is ususally necessary and always advantageous to have considerable information about a biosynthetic system prior to using ^{13}C labeled precursors. A number of studies on insect cuticular lipids have described the structures of the major hydrocarbon components (Nelson, 1978; Blomquist and Jackson, 1979). Subsequent studies with radioisotopes have suggested certain metabolic pathways and have given information regarding the relative rates of incorporation of a number of precursors (Jackson and Baker, 1970; Conrad and Jackson, 1971, Blomquist and Jackson, 1979; Blomquist et al., 1975; 1979b; Chu and Blomquist, 1980a; 1980b; Nelson, 1969; Blomquist and Kearny, 1976).

Cuticular hydrocarbons are relatively stable metabolic end products produced in reasonable abundance. This greatly enhances the suitablility of ^{13}C-NMR for locating labeled atoms. In several studies where cuticular hydrocarbon components were enriched from 2 to 50 fold from ^{13}C labeled acetate, propionate, succinate, and methylmalonate, no detectable label was observed in other lipid fractions, such as the triacylglycerols and phospholipids (Blomquist, et al., unpublished observations). The large amount of unlabeled material in these fractions undoubtedly resulted in greater dilution. This coupled with the metabolic turnover of the triacylglycerols and phospholipids make studies in which labeled precursors are administered over relatively long time periods less advantageous.

The method of introduction of ^{13}C labeled precursors into an organism for in vivo studies is critical. To achieve a high isotopic enrichment, the precursor must be available to the biosynbthetic system in sufficient concentration to compete effectively with the endogenous supply. This requirement was met in a study on the bio-synthesis is of 3-methylpentacosane in the cockroach, Periplaneta americana, by injecting 0.5 mg of substrate per insect each day over a 30 day time period. In smaller insects, such as termites and houseflies, in which repeated injections would cause high mortality, labeled substrates can be included in the diet (Blomquist et al., 1980a; Dillwith et al., 1981b). Prior knowledge of the insect's dietary preference is important however. For example, whereas the housefly readily consumes a diet containing up to one-fourth (by weight) of ^{13}C labeled sodium acetate or sodium propionate, the termite Zooter-mopsis angusticollis, the cricket Acheta domesticus, and the cockroach P. americana do not consume dietary material containing either of these substrates (Blomquist et al., unpublished observations).

A knowledge of when high rates of synthesis for a particular substance occur can also be most helpful. For example, the female hosefly synthesizes methylalkanes at a very low rate until about 2 days post-emergence, at which time high rates of synthesis are initiated (Dill-with et al., 1981a). When sodium [1-^{13}C] propionate was fed to 2 day old female housflies until day 4 or 5, a fifty fold enrichment in the carbon labeled by this substrate was obtained (Dillwith et al., 1981b).

NMR INSTRUMENT REQUIREMENTS

The natural abundance $\{^{1}H\}$ (proton decoupled) ^{13}C-NMR spectra included in Figure 1 were obtained with 2-14 mg of sample in 50 µl deuterated chloroform in a 1 mm micro-probe on a 100 MHZ instrument. Acceptable spectra were obtained on as little as 100 µg of significantly ^{13}C enriched material. Larger amounts of material are needed for samples in which the carbon containing ^{13}C is bound to a deuterium and a longer pulse delay time is required. The time needed to obtain spectra with a signal to noise ratio of greater than 50 was from 3 to 6 hours. The time required to obtain acceptable spectra is inversely pro-portional to the square of the sample concentration. Higher field strengths also decrease accumulation time significantly, but 1 mm probes are less common on instru-ments with higher field strengths.

ASSIGNMENT OF ^{13}C RESONANCES

Accurate interpretation of biosynthetic data from ^{13}C experiments requires an unequivocal assignment of all resonances. Discussions of methods for assigning resonances in ^{13}C-NMR spectra are available, and empir-ical rules for assigning resonances in the spectra of hy-drocarbons have been summarized (Levy and Nelson, 1972;

Strothers, 1976). The presence of attached and nearby carbons has a profound effect upon ^{13}C-NMR chemical shifts. Grant and Paul (1964) found that for a large number of simple hydrocarbons, each carbon gave a distinct chemical shift which was dependent upon the number of α-, β-, γ-, δ- and ϵ-carbon atoms. These empirical correlations allow an investigator to accurately assign the chemical shifts for each carbon, and are described in detail by Levy and Nelson (1972) and Stothers (1976). Using these empirical rules and associated chemical shift parameters, Dwyer et al. (1981a) found that the calculated chemical shifts for each carbon of 3-methylpentacosane agreed remarkably well with experimental values. Pomonis (unpublished data) has obtained the calculated and experimental chemical shift data for a series of long chain mono- and dimethylalkanes, and his data demonstrate that the empirical calculations described by Levy and Nelson (1972) and Stothers (1976) for small paraffins work very well for the long chain components found in insect waxes.

In a carbon-13 proton decoupled NMR spectrum each chemically non-equivalent carbon gives rise to a unique signal. For example, the ^{13}C-{^{1}H}-NMR spectrum of 3-methylpentacosane, a branched alkane from the cockroach, Periplaneta americana (Jackson, 1972) (Fig. 1A), gave eleven distinct resonances, with carbons 1-6, 23-25 and the branching methyl group having distinct signals (Dwyer et al., 1981a). The large signal at 29.8 ppm arises from the chemically equivalent carbons 7-22. Three components comprise approximately 90% of the branched alkanes of the termite Zootermopsis angusticollis: 5-methylheneicosane, a symmetrical 5,17-dimethylheneicosane, and 5-methyltricosane (Blomquist et al., 1979a). The ^{13}C-NMR spectrum (Fig. 1B) of a natural abundance mixture of these alkanes was almost identical to that of a 5-methylalkane standard. Carbons 2-7 and the methyl branch carbon of the monomethylalkanes had the same chemical shift as did carbons 2-7 from either

FIGURE 1 opposite. 25.00 MHz ^{13}C-{^{1}H} NMR spectra of CDCl$_3$ solutions containing (A) 3-methylpentacosane from P. americana. The inset shows the numbering system used for 3-methylpentacosane. (B) A mixture of methylalkanes from Z. angusticollis. The inset shows a gasliquid chromatography trace of the methyl branched alkanes. Components are (1) 5-methylheneicosane, (2) 5,17-dimethylheneicosane and (3) 5-methyltricosane. (C) The methyl branched alkanes from 4 day old female houseflies, M. domestica. The inset shows a gasliquid chromatography trace of branched alkanes which contain over 100 isomers and homologs. The numbers indicate where an n-alkane of that chain length would elute.

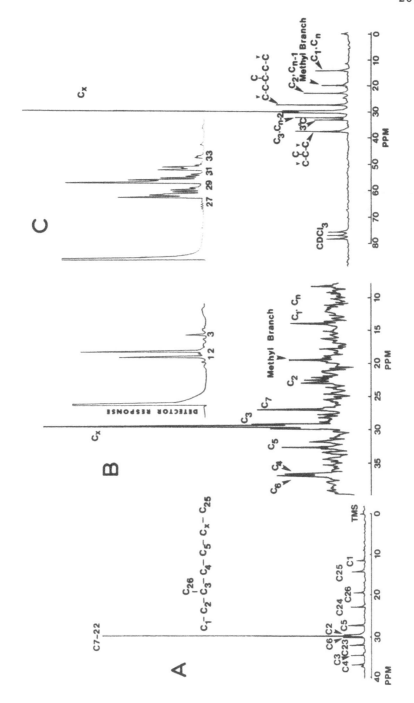

end of the symmetrical dimethylakane. Thus, the similarity in the chemical shifts of carbons in similar environments allowed meaningful data to be obtained, even though
the sample contained a mixture of alkanes. An even more
dramatic example of this occurred with the branched hydrocarbons of the housfly, in which, although there are
over 100 different isomers and homologs in the methylal-
kane fraction (Nelson et al., 1981), distinct signals
were observed for carbons 1 + n, 2 + n-1, 3 + n-2, 4 +
n-3, the methyl branch carbon, the tertiary carbon, and
the carbon adjacent to and once removed from the tertiary
carbon (Fig. 1C). The methyl branch shifts the signal
from nearby carbons and results in each carbon near the
methyl branch giving distinct signals.

BIOSYNTHESIS OF n-ALKANES

In the past few years the pathways leading to the
biosynthesis of the major insect cuticular lipids have
been established. Much of this information has been
reviewed elsewhere (Blomquist and Jackson, 1979; Howard
and Blomquist, 1982) and only a summary will be presented here.
Two major routes for the biosynthesis of hydrocarbons have been considered: (1) the condensation-reduction and (2) the elongation-decarboxylation pathways
(Fig. 2). In the condensation-reduction pathway, two
molecules of fatty acids condense head-to-head with
decarboxylation of one of them, to give a symmetrical
ketone which is subsequently reduced to a hydrocarbon.
There is evidence supporting the existence of this
pathway in bacteria (Albro, 1976). Plants (Kolattukudy
et al., 1976) and insects (Blomquist and Jackson 1979),
however, appear to utilize the elongation-decarboxylation pathway. Here, a C_{16} or C_{18} fatty acid is
elongated to a very long chain fatty acid which is then
decarboxylated to yield an alkane one carbon unit
shorter. Direct evidence for the elongation-decarboxylation pathway in insects was obtained by demonstrating in
vivo the direct decarboxylation of [R-^{3}H]hexacosanoic
acid to n-pentacosane in P. americana (Major and Blomquist, 1978) and of [15,16-^{3}H]tetracosanoic acid to n-
tricosane both in vivo and in vitro in Z. angusticollis
(Chu and Blomquist, 1980a).
Examination of the ^{13}C-NMR spectrum of 3-methylpen-
tacosane in P. americana enriched from [1-^{13}C] acetate
and [2-^{13}C] acetate gave additional evidence consistent
with the elongation-decarboxylation pathway (Dwyer et al.
1981a). The incorporation of the label from [1-^{13}C]
acetate into carbons 2, 6 and 24 and from [2-^{13}C] acetate
into carbons 1, 5, 23 and 25 of 3-methylpentacosane in
P. americana to approximately an equal extent is consis-

tent with an elongation-decarboxylation pathway for methylalkanes, although, by itself, it does not rule out the possibility of alternative routes. It is not possible to ascertain whether carbons 8-22 are labeled, since similar chemical environments result in a common signal from these carbons. Presumably the odd numbered carbons are labeled from [2-^{13}C] acetate and the even numbered carbons from [1-^{13}C] acetate.

BRANCHED ALKANES

The biosynthesis of methyl alkanes has received considerable attention since they often comprise a large percentage of the hydrocarbon fraction and function both as pheromones and kairomones in many insect species (Blomquist and Jackson, 1979).
    In birds, the methyl branches of multiple methyl branched fatty acids arise from the incorporation of propionate, as the methylmalonyl-CoA derivative, in place of malonyl-CoA during chain elongation (Buckner et al.,

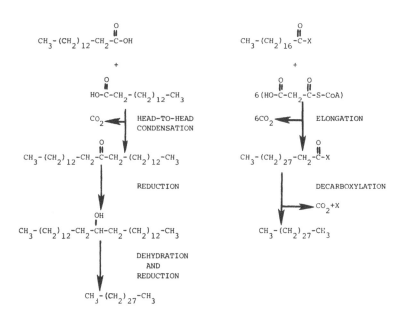

FIGURE 2. Proposed pathways for the biosynthesis of hydrocarbons. Left: Condensation-reduction pathway; Right: Elongation-decarboxylation pathway.

1976). Plants (Kolattukudy et al., 1976), microorganisms
(Albro, 1976), and insects (Blailock et al., 1976) appear
to utilize the carbon skeletons of valine and leucine as
"starter units" in the biosynthesis of 2-methylalkanes.
The carbon skeleton of isoleucine serves as the "starter
unit" for 3-methylalkanes in plants (Kolattukudy et al.,
1976). In contrast, evidence has accumulated indicating
that the biosynthesis of 3-methylalkanes and internally
branched alkanes (Blomquist and Jackson, 1979) in insects occurs via the incorporation of propionate (as the
methylmalonate derivative) in place of malonate, and the
carbon skeleton of amino acids is not incorporated as an
intact unit.

Studies on the biosynthesis of 3-methylpentacosane
in the cockroach P. americana indicated that propionate
was the precursor to the methyl branch unit. Radio-GLC
showed that $[1-^{14}C]$propionate was selectively incorporated into 3-methylpentacosane. It was not possible
to determine unequivocally which carbon was labeled because degradative studies are difficult to perform on
normal and methyl branched alkanes. A knowledge of the
exact position of the label was necessary to determine
whether the methyl branching unit is incorporated during
the early or the late steps of chain elongation. Due to
the fact that ω-3 methyl branched fatty acids have not
been found in this insect (Dwyer, 1980), it was proposed
that the methyl branch unit was incorporated during the
penultimate step of chain elongation, although direct
evidence was not presented.

This hypothesis was tested using carbon-13 labeled
precursors and examining the hydrocarbons by ^{13}C-NMR.
If carbon 2 of 3-methylpentacosane was enriched from
$[1-^{13}C]$ propionate, the methyl branch would have been
incorporated during the penultimate step of chain elongation. Alternatively, if carbon 4 were enriched, the
methyl branch would have been added as the second unit
of the chain (Fig. 3). The ^{13}C-NMR spectrum of the
labeled alkane enriched from $[1-^{13}C]$ propionate shows
that all of the label was incorporated into position 4
(Fig. 3), indicating that the methyl branch unit was
added as the second unit during chain synthesis (Dwyer
et al., 1981a). When 3-methylpentacosane was enriched
from $[methyl-^{13}C]$methylmalonate, the label was preferentially incorporated into the branching methyl carbon,
suggesting that propionate was converted to a methylmalonyl derivative prior to incorporation. Mass spectrometry was also used to examine the incorporation of
$[1-^{13}C]$ propionate into 3-methylpentacosane. The mass
spectral results were consistent with the NMR data, but,
by themselves, were not definitive.

Indirect evidence from studies with ^{14}C and ^3H
labeled precursors (Chu and Blomquist, 1980b) suggested

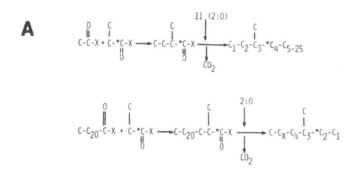

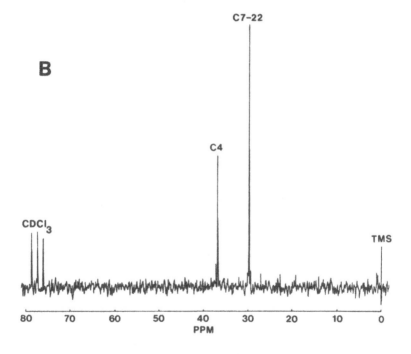

FIGURE 3. Biosynthesis of 3-methylpentacosane. (A) Two possible pathways for the incorporation of [1-^{13}C] propionate into 3-methylpentacosane (B) 25.00 MHz ^{13}C-{^{1}H} NMR spectrum of a $CDCl_3$ solution containing 4 mg 3-methylpentacosane enriched from [1-^{13}C] propionate. This spectrum indicates that only carbon 4 is labeled.

that in the termite Z. angusticollis, in additon to propionate, succinate could be converted to a methylmalonyl-
derivative, which was then incorporated into the alkyl
chain as the branching unit. Direct evidence for this
was obtained from ^{13}C-NMR studies (Fig. 4). The labeled carbons from [2,3-^{13}C] succinate were converted
to the methyl branch and teriary carbon of the branched
alkanes (Blomquist et al., 1980a). When a substrate labeled with ^{13}C on adjacent carbons is used, its incorporation without cleavage of the bond between the labeled carbons is readily ascertained by the resulting
^{13}C-^{13}C coupling in the labeled product (Fig. 4).
The signals 17.5 Hz to either side of the signals from
the methyl branch and tertiary carbon (Fig. 4) arise
from ^{13}C-^{13}C coupling of the adjacent carbons. In
spectra from singly labeled and natural abundance
samples, ^{13}C-^{13}C coupling is not observed because of
the low natural abundance of ^{13}C (1.1%).

The studies discussed above clearly indicate that
propionate and succinate, as the methylmalonyl derivative, serve as the source of the methyl branch unit in
methyl branched alkanes of the termite. However, the
question of which one is the physiological precursor to
methylmalonyl-CoA is still unanswered.

In the housefly, Musca domestica L., cuticular hydrocarbons serve as sex pheromone components. Carlson et
al. (1971) identified (Z)-9-tricosene as a major component of the pheromone, and a series of methyl branched alkanes (Nelson et al., 1981) were shown to potentiate the
activity of Z-(9)-tricosene (Uebel et al., 1976; Rogoff
et al., 1980). The biosynthesis of the hydrocarbon components was studied using radioactive substrates (Dillwith et al., 1981a). The data indicated that alkenes
were formed by the desaturation of stearate to oleate,
which is then elongated and decarboxylated. [1-^{14}C]
Propionate preferentially labeled the methylakanes, but
it was not possible to determine if succinate contributed to the formation of methylalkanes, and the difference in the metabolism of propionate labeled in the 2
and 3 positions versus the 1 position raised questions
regarding the metabolic fate of this precursor.

Carbon-13-NMR spectroscopy was used to answer some
of the questions raised by the studies with radioactive
precursors. The labeled carbon from [1-^{13}C] acetate
was incorporated into the even numbered carbons of the
alkanes and alkenes, and [2-^{13}C] acetate labeled the
odd numbered carbons. The labeled carbons from [3-^{13}C]
propionate and [methyl-^{13}C] methylmalonate were incorporated into the methyl branch carbon of the methylalkanes, and the labeled carbons from [2-^{13}C] and
[1-^{13}C] propionate labeled the tertiary carbon and carbon adjacent to the tertiary carbon, respectively (Dill-

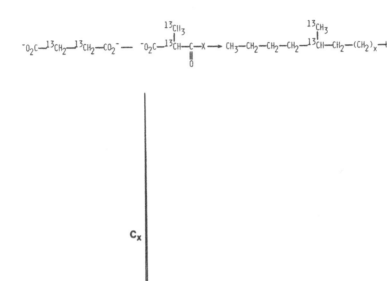

FIGURE 4. Incorporation of [2,3-^{13}C] succinate into methylalkanes. (A) Proposed pathway, (B) 25.00 MHz ^{13}C-{^{1}H}NMR spectrum, of a CDCl$_3$ solution of methylalkanes enriched from [2,3-^{13}C] succinate showing ^{13}C-^{13}C coupling. This spectrum indicates that the methyl branch and the tertiary carbon are labeled.

214

with et al., 1981b). There was no incorporation of the label from $[2,3-^{13}C]$ succinate into the methyl branch or tertiary carbon, but carbons 1 + n, 2 + n-1, and 3 + n-2 were enriched. This demonstrates that succinate does not contribute to the methyl branching unit in the housefly.

A proposed pathway for the biosynthesis of methylalkanes in the housefly is presented in Fig. 5, and shows the origin of each carbon in the branched alkanes. Preliminary data with radioactive valine and isoleucine suggest that amino acids which are metabolized via propionyl-CoA and methylmalonyl-CoA are the physiological precursors to the branching methyl group (Dillwith and Blomquist, unpublished data).

Studies on the incorporation of radiolabeled propionate into hydrocarbons of the housfly (Dillwith et al., 1981b) and other insects (Chu and Blomquist, 1980b; Blomquist et al., 1975; Kearney and Blomquist, 1976) and into juvenile hormone in Manduca sexta (Schooley et al., 1973) suggested that propionate was converted to acetate with the loss of the carboxyl carbon. Direct evidence for

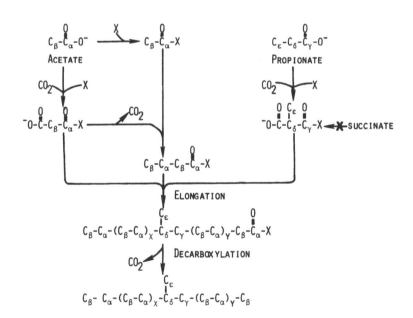

FIGURE 5. Summary pathway for methyl branched hydrocarbon biosynthesis in the housefly.

this was obtained by examining the metabolism of [2-^{13}C] and [3-^{13}C] propionate in the housefly. In addition to being incorporated as the methyl branch unit, [3-^{13}C] propionate also labeled the even numbered carbons, and [2-^{13}C] propionate labeled the odd numbered carbons of both the alkanes and alkenes. This suggests that propionate is converted to an acetyl derivative, with carbon 3 of propionate converted to the carboxyl carbon of acetate, and carbon 2 of propionate converted to the methyl carbon of acetate. This modified ß-oxidation of propionate has been shown to occur in plant tissue, and the mitochondrial fraction was most active in carrying out this conversion (Giovanelli and Stumpf, 1958; Hatch and Stumpf, 1962).

The occurrence of this rather unusual pathway for the metabolism of propionate prompted experiments to determine if the housefly and several other insects contained vitamin B12. This vitamin, which is absent in plants, is required for the interconversion of methylmalonyl-CoA and succinyl-CoA. The presence of vitamin B12 was not detected in the housefly (Dillwith, Wakayama, and Blomquist, unpublished). Thus, since the housefly cannot metabolize propionate via methylmalonyl-CoA to succinyl-CoA, it has apparently developed alternative routes for propionate metabolism. In contrast, the termite, Z. angusticollis, contained copious amounts of vitamin B12 both in its gut tract and body tissues. The termite can convert succinate to a methylmalonyl derivative (Blomquist, et al., 1980a), and presumably can convert propionate to succinate.

DUAL LABEL STUDIES

An attempt was made to obtain information on the mechanism of the decarboxylation of long chain fatty acids to alkanes in insects by the use of a double labeled [2-^{13}C, 2,2,2-^{2}H] acetate in a manner similar to the studies done with n-heptadecane in the blue-green alga Anacystis nidulans (McInnes et al., 1980). When a ^{13}C atom is bound to a deuterium, the signal from the ^{13}C atom is shifted and split in proton decoupled spectra, thus yielding information about the number of deuteriums directly linked to the ^{13}C atom. The signals from C-1 and C-25 are distinct in 3-methylpentacosane from P. americana, a fact which allows direct observation of the two ends of the molecule. Although 3-methylpentacosane was readily enriched at odd numbered carbons in ^{13}C from [2-^{13}C, 2,2,2-^{2}H] acetate, no retention of the deuterium was observed. Similar data were obtained in the female housefly. However, both ^{13}C and ^{2}H were incorporated into the n-alkanes and alkenes of the male housefly, which does not synthesize significant amounts of methyl-branched hydrocarbons (Blomquist and Dillwith, unpublished observations). The fact that the deuterium label is lost in the cockroach and female housefly, but retained in the male housefly, suggests that acetate is directly incorporated into n-alkanes and n-alkenes in the male housefly, but undergoes metabolic changes in the other cases. Since a high level of methylmalonyl-CoA may be needed to synthesize branched hydrocarbons in epidermal tissue, an active malonyl-CoA decarboxylase, which converts malonyl-CoA to acetyl-CoA, may be present. This would result in a cycling of malonyl-CoA to acetyl-CoA and back to malonyl-CoA, and could result in the loss of the deuteriums from acetate. This situation could be analgous to that found in the uropygial gland of birds, where the synthesis of multiple methyl branched fatty acids is due to high levels of methylmalonyl-CoA brought about by an active malonyl-CoA decarboxylase (Buckner et al., 1976). Experiments are now being conducted to further investigate this possiblity in insects.

UNSATURATED HYDROCARBONS

The biosynthesis of mono- and diunsaturated hydrocarbons has been investigated in several insect species, and ^{13}C-NMR has been used in studies on the biosynthesis of the alkenes of the housefly (Dillwith et al., 1981b) and an alkadiene in the American cockroach (Dwyer, et al., 1981b).

Alkadienes are not common insect cuticular lipid components, but an exception, (Z,Z)-6,9-heptacosadiene, comprises about 70% of the cuticular hydrocarbon of P. americana (Baker et al., 1963; Jackson, 1972). The similarity in the positions of the double bonds in the C_{27} alkadiene compared to linoleic acid suggested that linoleate could be elongated and decarboxylated (Jackson and Baker, 1970). Recent studies have demonstrated that [1-^{14}C] acetate is incorporated about equally into the alkane and alkadiene fractions, whereas [1-^{14}C] linoleate preferentially labelled the alkadiene, and [9,10-^{3}H] oleate labeled the alkadiene almost exclusively (Dwyer et al., 1981b). Both [1-^{14}C] acetate and [9,10-^{3}H] oleate were readily converted to linoleate in this insect (Dwyer, 1980). This suggested that linoleate was synthesized de novo, and then elongated and decarboxylated to form (Z,Z)-6,9-heptacosadiene. However, the ^{13}C-NMR spectrum of the diene labeled from [2-^{13}C]acetate showed that only carbons 25 and 27 were labeled. Ozonolysis of the [1-^{14}C] acetate labeled diene followed by radio-GLC confirmed that carbons 1-6 were not labeled, whereas the fragment containing carbons 10-27 was labeled. A hypothesis which could explain the radio and stable isotope data is as follows: linoleic acid is synthesized in the American cockroach (Dwyer, 1980) and is stored in the insect. Since P. americana contains a large amount of linoleate, the diunsaturated fatty acid labeled from $[1-^{14}C]$ or $[2-^{13}C]$ acetate is diluted. Thus, as alkadiene synthesis proceeds, the cockroach uses linoleate from its stores, and because of isotopic dilution, no detectable radioactivity from acetate would be present in carbons 1-18 of the C_{27} diene. Indirect evidence supporting this hypothesis is the observation that neither oleate nor linoleate contained detectable label from $[1-^{13}C]$ or $[2-^{13}C]$ acetate, even in experiments where the hydrocarbon was readily labeled (Bloomquist et al., unpublished). Dilution of the ^{13}C label by large amounts of stored lipid and rapid turnover of the acyl moieties could account for the lack of ^{13}C label in the fatty acids.

In contrast to the situation in the cockroach, the biosynthesis of the alkanes in the honeybee appears to occur in a tightly coupled complex. In the cuticular and comb wax of the honeybee, Apis mellifera, the major isomer of the shorter chain $C_{23}-C_{29}$ minor components is the (Z)-9 isomer (Blomquist, et al., 1980c). However, the C_{31} component is comprised of about equal amounts of (Z)-8 and (Z)-10 isomers, and the C_{33} component is almost exclusively (Z)-10 (Blomquist et al., 1980c). Labeled oleic acid was incorporated only about one-tenth as efficiently into the hydrocarbons of the honey bee as was stearate, and was incorporated primarily into minor components, the $C_{23}-C_{29}$ (Z)-9 alkenes. In contrast, $[1-^{14}C]$ acetate and $[1-^{14}C]$ stearate primarily labeled the (Z)-8 and (Z)-10 C_{31} and C_{33} components. These data suggest that once desaturation occurs at the $\omega-8$ and $\omega-10$ positions, the alkenoic acid is then elongated and decarboxylated without being released from the enzyme complex (Blomquist et al., 1980b). In accord with this, an examination of the C_{18} monosaturated fatty acids of the honeybee showed that the double bond was exclusively in the (Z)-9 position.

CONCLUDING COMMENTS

The information discussed in this chapter demonstrates the usefulness of ^{13}C-NMR spectroscopy in biosynthetic studies on insect cuticular hydrocarbons. Carbon-13-NMR facilitates the determination of the exact location of a labeled carbon without degradative studies, and in many cases, allows definitive statements to be made regarding metabolic pathways. It has been used to determine the origin of all the carbons in a number of insect hydrocarbons, and to determine the source of the methyl branching unit of methylalkanes.

The biosynthesis of the methyl branched hydrocarbon

in insects appears to occur in a tightly coupled elongation-decarboxylation system. Methyl branched fatty acids
comprise at most a minor portion of most insect lipids,
suggesting that once the methyl branch is inserted, the
molecule is not released from the enzyme system until it
is decarboxylated. The approximately equal incorporation
of ^{13}C from ^{13}C labeled acetate into both ends of
methylalkanes (Dwyer et al., 1981a) is consistent with a
coupled system. It appears that alkene sysnthesis can
proceed either by the elongation of pre-formed unsaturated fatty acids (Dwyer et al., 1981), or in a complex,
in which once desaturation occurs, the molecule is directly elongated and decarboxylated (Blomquist et al.,
1980).

It is likely that ^{13}C-NMR will be extensifely used
in future studies on a variety of lipids and other biomo-
lecules in insects. The use of ^{13}C-NMR to monitor the
metabolism of ^{13}C labelled substrate with in vitro systems will allow increasingly sophisticated questions to
be answered with this technique. The use of precursors
with ^{13}C on adjacent carbons and precursors containing
deuterium attached to ^{13}C atoms can increase the amount
of information that can be obtained from ^{13}C-NMR studies.
The major limitiations in this work are the necessity of
reasonably large sample sizes and the need for significant isotopic enrichment.

ACKNOWLEDGEMENT

The recent work from our laboratory was supported in
part by the Science and Education Administration of the
U.S. Department of Agriculture under grant 7801064 from
the Competitive Research Grants Office. Contribution of
the Nevada Agricultural Experiment Station, Journal
Series No. 531. The authors thank Dr. John H. Nelson,
Department of Chemistry, University of Nevada, Reno,
89557, for his suggestions on this manuscript and for
stimulating interactions that contributed importantly to
ideas expressed here. We thank Dr. Ralph W. Howard,
U.S.D.A. Forest Service, Forestry Science Laboratory,
P.O. Box 2008, GMF, Gulfport, MS., for his suggestions
in the preparation of this manuscript.

REFERENCES

Albro, P. W. (1976) Bacterial waxes. In Chemistry and
    Biochemistry of Natural Waxes. ed. P.E. Kolattukudy,
    pp 419-445, Amsterdam, Elsevier.
Baker, G. L., Vroman, H. E. and Padmore, J. (1963) Hy-
    drocarbons of the American cockroach. Biochem. Bio-
    phys. Res. Comm. 13: 360-365.
Blailock, T.T., Blomquist, G.J. and Jackson, L.L. (1976)

Biosynthesis of 2-methylalkanes in the crickets <u>Nemobius</u> <u>fasciatus</u> and <u>Gryllus</u> <u>pennsylvanicus</u>. <u>Biochem</u>. <u>Biophys</u>. <u>Res</u>. <u>Comm</u>. 68: 841-849.

Blomquist, G.J., Chu, A.J. and Nelson, J.H. and Pomonis, J.G. (1980a) Incorporation of [2,3-^{13}C] succinate into methyl branched alkanes in a termite. <u>Arch</u>. <u>Biochem</u>. <u>Biophys</u>. 204: 648-650.

Blomquist, G.J., Chu, A.J. and Remaley, S. (1980b) Biosynthesis of wax in the honeybee, <u>Apis</u> <u>mellifera</u> L. <u>Insect</u> <u>Biochem</u>. 10: 313-321.

Blomquist, G.J., Howard, R.W., McDaniel, C.A., Remaley, S., Dwyer, L.A. and Nelson, D.R. (1980c) Application of mercuration-demercuration followed by mass spectometry as a convenient microanalytical technique for double bond location in insect derived alkenes. <u>J</u>. <u>Chem</u>. <u>Ecol</u>. 6: 257-269.

Blomquist, G.J., Howard, R.W. and McDaniel, C.A. (1979a) Structure of the cuticular hydrocarbons of the termite <u>Zootermopsis</u> <u>angusticollis</u> (Hagen) <u>Insect</u> <u>Biochem</u>. 9: 365-370.

Blomquist, G.J., Howard, R.W. and McDaniel, C.A. (1979b) Biosynthesis of the cuticular hydrocarbons of the termite <u>Zootermopsis</u> <u>angusticollis</u> (Hagen). Incorporation of propionate into dimethylalkanes. <u>Insect</u> <u>Biochem</u>. 9: 371-374.

Blomquist, G.J. and Jackson, L.L. (1979) Chemistry and biochemistry of insect waxes. <u>Progress</u> <u>in</u> <u>Lipid</u> <u>Research</u> 17: 319-345.

Blomquist, G.J. and Kearney, G.P. (1976) Biosynthesis of the internally branched monomethylalkanes in the cockroach <u>Periplaneta</u> <u>fuliginosa</u>. <u>Arch</u>. <u>Biochem</u>. <u>Biophys</u>. 173: 546-553.

Blomquist, G.J., Major, M.A. and Lok, J.B. (1975) Biosynthesis of 3-methylpentacosane in the cockroach <u>Periplaneta</u> <u>americana</u>. <u>Biochem</u>. <u>Biophys</u>. <u>Res</u>. <u>Comm</u>. 64:43-50.

Buckner, J.S., Kolattukudy, P.E. and Poulose, A.J. (1976) Purification and properties of malonyl-coenzyme A decarboxylase, a regulatory enzyme from the uropygial gland of goose. <u>Arch</u>. <u>Biochem</u>. <u>Biophys</u>. 177:539-551.

Carlson, D.A., Mayer, M.S., Silhacek, D.L., James, J.D., Beroza, M. and Bierl, B.A. (1971) Sex Attractant pheromone of the housefly: Isolation, identification and synthesis. <u>Science</u> 174: 76-77.

Chu, A.J. and Blomquist, G.J. (1980a) Decarboxylation of tetracosanoic acid to <u>n</u>-tricosane in the termite <u>Zootermopsis</u> <u>angusticollis</u>. <u>Comp</u>. <u>Biochem</u>. <u>Physiol</u>. 66B: 313-317.

Chu, A.J. and Blomquist, G.J. (1980b) Biosynthesis of hydrocarbons in insects: Succinate is a precursor of the methyl branched alkanes. <u>Archiv</u>. <u>Biochem</u>. 201: 304-312.

Conrad, C.W. and Jackson, L.L. (1971) Hydrocarbon bio-
    synthesis in Periplaneta americana. J. Insect Physi-
    ol 17:1907-1916.
Dillwith, J.W., Blomquist, G.J. and Nelson, D.R. (1981a)
    Biosynthesis of the hydrocarbon components of the sex
    pheromone of the housefly, Musca domestica. Insect
    Biochem. 11: 247-253.
Dillwith, J.W., Nelson, J.H., Pomonis, J.G., Hakk, H. and
    Blomquist, G.J. (1981b) A ^{13}C-NMR study of the
    biosynthesis of the cuticular hydrocarbon and sex
    pheromone components in the housefly. Biochemistry
    (submitted).
Dwyer, L.A. (1980) Lipid biosynthesis in the American
    cockroach, Periplaneta americana. Ph.D. Disserta-
    tion, University of Nevada, Reno. 93 pp.
Dwyer, L.A. and Blomquist, G.J. (1981) Biosynthesis of
    linoleic acid in the cockroach, P. americana. Pro-
    ceedings of the 25th Jubilee Congress on Essential
    Fatty Acids and Prostaglandins. Prog. Lipid Res.
    20: 215-218.
Dwyer, L.A., Blomquist, G.J., Nelson, J.H. and Pomonis,
    J.G. (1981a) A ^{13}C NMR study of the biosynthesis
    of 3-methylpentacosane in the American cockroach.
    Biochem. Biophys. Acta 663: 536-544.
Dwyer, L.A., de Renobales, M. and Blomquist, G.J. (1981b)
    Biosynthesis of (Z,Z)-6,9-heptacosadiene in the
    American cockroach. Lipids (submitted).
Giovanelli, J. and Stumpf, P. K. (1958) Fat metabolism
    in plants. X Modified ß-oxidation of propionate by
    peanut mitochondria. J. Biol. Chem. 231: 411-426.
Grant, D. M. and Paul, E. G. (1964) Carbon-13 magnetic
    resonance. II. Chemical shift data for the alkanes.
    J. Am. Chem. Soc. 86: 2984-2990.
Grutzner, J. B. (1972) Carbon-13 NRM spectroscopy and
    its application to biological systems. Lloydia 35:
    375-398.
Hadley, N.F. (1980) Surface waxes and integumentary
    permeability. Amer. Scientist 68: 546-553.
Hatch, M.D. and Stumpf, P.K. (1962) Fat metabolism in
    higher plants. XVIII. Propionate metabolism by plant
    tissues. Arch. Biochem. Biophys. 96: 193-198.
Howard, R.W. and Blomquist, G.J. (1982) Chemical ecology
    and biochemistry of insect hydrocarbons. Ann. Rev.
    Ent. (in press).
Howard, R.W., McDaniel, C.A., Nelson, D.R., Blomquist,
    G.J., Gelbaum, L.T. and Zalkow, L.H. Cuticular hy-
    drocarbons as possible species- and caste-recognition
    cues in Reticulitermes sp. J. Chem. Ecol. (in prep-
    aration).
Jackson, L.L. (1972) Cuticular lipids of insects IV.
    Hydrocarbons of the cockroaches Periplaneta japonica
    and Periplaneta americana compared to other cockroach

hydrocarbons. Comp. Biochem. Physiol. 41B: 331-336.
Jackson, L.L. and Baker, G.L. (1970) Cuticular lipids of insects. Lipids 5: 239-246.
Jackson, L.L. and Blomquist, G.J. (1976) Insect waxes. In Chemistry and Biochemistry of Natural Waxes. ed. P.E. Kolattukudy, Elsevier, Amsterdam pp. 201-233.
Kolattukudy, P.E., Croteau, R. and Buckner, J.S. (1976) Biochemistry of plant waxes. In Chemistry and Biochemistry of Natural Waxes. ed. P.E. Kolattukudy Elsevier, Amsterdam pp. 298-347.
Kunesch, G. and Poupat, C. (1977) Biosynthetic studies using carbon-13 enriched precursors. In Isotopes in Organic Chemistry, Buncel, E. and Lee, C.C. ed. Elsevier, Amsterdam pp. 105-170.
Levy, G.C. and Nelson, G.L. (1972) Carbon-13 Nuclear Magnetic Resonance for Organic Chemists. Wiley-Interscience, New York, pp. 38-78.
Lockey, K.H. (1980) Insect cuticular hydrocarbons. Comp. Biochem. Physiol. 65B: 457-462.
London, R.E., Kollan, V.H. and Matwiyoff, N.A. (1975) The quantitative analysis of carbon-carbon coupling in the ^{13}C nuclear magnetic resonance spectra of molecules biosynthesized from ^{13}C enriched precursors. J. Amer. Chem. Soc. 97: 3565-3573.
Major, M.A. and Blomquist, G.J. (1978) Biosynthesis of hydrocarbons in insects: Decarboxylation of long chain acids to n-alkanes in Periplaneta. Lipids. 13: 323-328.
McInnes, A.G., Walter, J.A. and Wright, J.L.C. (1980) Biosynthesis of hydrocarbons by algae: Decarboxylation of stearic acid to n-heptadecane in Anacystis nidulans determined by ^{13}C- and ^{2}H-labeling and ^{13}C nuclear magnetic resonance. Lipids 15: 609-615.
McInnes, A.G., Walter, J.A., Wright, J.L.C., and Vining, L.C. (1976) ^{13}C NMR biosynthetic studies. In Topics in carbon-13 NMR spectroscopy. Levy, G.C., ed. John Wiley and Sons, New York. pp. 123-178.
Nelson, D.R. (1969) Hydrocarbon synthesis in the American cockroach. Nature Lond. 221: 854-855.
Nelson, D.R. (1978) Long-chain methyl-branched hydrocarbons: occurrence, biosynthesis and function. Adv. Insect Physiol. 13: 1-33.
Nelson, D.R., Dillwith, J.W. and Blomquist, G.J. (1981) Cuticular hydrocarbons of housefly Musca domestica. Insect Biochem. 11: 187-197.
Rogoff, W.M., Gretz, G.H., Sonnet, P.E. and Schwarz, M. (1980) Responses of male houseflies to muscalure and to combinations of hydrocarbons with and without muscalure. Environ. Entomol. 9: 605-606.
Schooley, D.A., Judy, K.J., Bergot, B.J., Hall, M.S. and Siddall, J.B. (1973) Biosynthesis of the juvenile hormones of Manduca sexta: labeling pattern from

mevalonate, propionate and acetate. <u>Proc</u>. <u>Natl</u>. <u>Acad</u>. <u>Sci</u>. <u>USA</u> 70:2921-2925.

Sequin, V. and Scott, A.I. (1974) Carbon-13 as a label in biosynthetic studies. <u>Science</u> 186: 101-107.

Simpson, T.J. (1975) Carbon-13 nuclear magnetic resonance in biosynthetic studies. <u>Chemical</u> <u>Society</u> <u>Reviews</u> 497-522.

Stothers, J.B. (1976) Carbon-13-NMR spectroscopy. Academic Press, New York. pp. 102-127.

Uebel, E.C., Sonnet, P.E. and Miller, R.W. (1976) Housefly sex pheromone: Enhancement of mating strike activity by combination of (Z)-9-tricosene with branched saturated hydrocarbons. <u>J</u>. <u>Econ</u>. <u>Entom</u>. 5: 905-908.

12
Lipid Nutrition and Metabolism of Cultured Insect Cells

James L. Vaughn
and Spiro J. Louloudes*

INTRODUCTION

During the years, 1950-1960, there was a great deal of research that demonstrated the importance of lipids to insects and particularly their requirements for exogenous sources of sterols and the fatty acids, linoleic and linolenic (Chippendale, 1972). It was during this same time that much of the early development of insect cell cultures and cell culture media took place, yet the role of lipids in the nutrition of insect cells in culture was virtually ignored. Grace (1958) tested cholesterol in primary cultures of silkworm cells and reported that it had no effect on cell migration, division or survival and for several years afterwards no further attention was paid to the requirements of insect cells for lipids. Choline chloride was added to some media but no studies were made to determine its importance until it was shown by Nagle (1969) that it was essential for optimum cell growth.

In retrospect, two factors probably account for this lack of interest in the role of lipids in cell nutrition. First the culture media were formulated with the assumption that either insect hemolymph or serum from other animals was required (Vaughn, 1971). These various materials contained more than enough lipids to meet at least the minimal requirements of the culture systems in use. So there was no apparent need for specific lipids in the culture media (Vaughn et al., 1971, and Louloudes et al., 1973).

The second obstacle which may have inhibited the use of lipid supplements was the problem associated with the incorporation of these materials into the aqueous media. However, as the interest and the need developed to reduce

* Plant Protection Institute, U.S. Department of Agriculture, Beltsville, MD 20705, USA the amounts of the natural, but undefined and highly variable serum supplements with more defined and less variable media components, methods of incorporating lipids into culture media were tested. Three different methods are commonly used and before discussing the metabolism and the nutritional requirements of lipids themselves it would be useful to discuss these methods.

One widely used approach is to dissolve the desired lipids in ethanol and Tween 80 and sterilize the solution by filtration. The sterile solution is then mixed with culture media. This procedure must be used with care as some cells are sensitive to either ethanol or Tween 80 (Brooks et al., 1980) or the Tween 80 itself may be a cell growth stimulant (Vaughn and Louloudes, 1972).

To avoid these problems some investigators have used various proteins as carriers for lipids. The protein is first extracted with organic solvents to remove the naturally occurring lipids (Rothblat et al. 1976). In addition to the bovine serum albumin used by Rothblat et al. whole serum has been used (Louloudes et al., 1973). The delipidization is not only a lengthy process but the test lipids may not bind to the protein in a manner which makes them available to the cell. Thus false negative results may be obtained.

The use of liposomes as vehicles for the cellular incorporation of biologically active materials has been applied in a variety of situations (Paphadjpopoulos et al., 1980). Phospholipid vesicles are prepared which enclose an aqueous solution containing the test materials. During incubation with the cells in culture the liposomes are taken into the cell without cytotoxic effects. This method is one in which defined materials can be used exclusively, therefore, the interpretation of the results is less difficult.

The method used to incorporate the lipid materials does not appear to have influenced the results except in one or two studies which will be discussed in the appropriate place. Instead, the study of lipids in cell culture is limited by the availability of cell lines. Most of the insect cell lines have originated from either Lepidoptera or Diptera and the studies are limited to a cell line from a few species in these two orders plus the order Orthoptera. Also, limited tissues are represented by the available cell lines. Most cell lines from Lepidoptera are from ovaries and most cell lines from Diptera are embryonic or from minced neonate larvae, i.e. unknown tissue sources. No cell lines have been established from fat body tissues which would be most useful in the study of lipids. It is important to keep in mind these limitations and to avoid making generalizations from what must be considered a limited sampling.

FATTY ACIDS

The role of fatty acids in insect nutrition, their transport and their metabolism in insects is of great interest and has been widely studied as evidenced by the papers by Dadd, Thompson and Turunen in this volume. A similar interest exists for the study of these phenomena in cell culture. In addition, the role that fatty acids have in determining the characteristics of the phospho-lipids in cell membranes and thus the characteristics of these membranes has been of interest to the virologist studying the replication of arboviruses in their insect vectors.

Insect cells, and most other animal cells, are main-tained in culture in a liquid medium composed of a mix-ture of inorganic salts, amino acids, sugars, vitamins and occasionally other defined chemicals. This mixture is further supplemented with one or more complex natur-ally occurring materials such as various animal sera. The most common of these are bovine sera, generally fetal bovine serum, and occasionally newborn calf serum. These sera are the sources of the fatty acids needed for insect cell growth. One such medium used to culture cells from several species of mosquitoes was supplemented with 20% newborn calf serum and in this medium 56.5% of the total lipids present were neutral lipids (McMeans et al., 1975). Free fatty acids comprised 15.1% and triglycer-ides comprised 19.2% of the neutral lipids. Table 1 shows the distribution of fatty acids within the total lipid fraction of the medium and the distribution of fatty acids in the neutral lipid fraction of the cells of several species of mosquitoes during the log phase of growth. These data indicate that the growing cells met-abolized the fatty acids and did not merely incorporate fatty acids from the medium, e.g. the increased cellular levels of 20:0, 20:5, and 16:1 and a lower amount of linoleic (18:2) than were present in the medium. Cells from all species of mosquitoes seemed to be capable of desaturation and chain elongation of fatty acids. The authors also determined the levels of fatty acids in the cells when the culture had reached the stationary phase and growth had ceased (McMeans et al., 1975; McMeans et al., 1976). They found the fatty acids of all 4 species showed increases in the chain length and/or in the amount of desaturation in the non-growing cells. The cells con-tained less oleic (18:1) and more linoleic (18:2) than did actively growing cells.

The distribution of fatty acids was further studied in the cells of the two Aedes species (Yang et al., 1976a). Only 5.0% of the neutral lipids in the A. aegypti were found to be free fatty acids and only 3.6% in A. albopictus as compared to 15.1% in the medium. In

TABLE 1.

Profiles of the principal fatty acids from total lipids of 4 species of mosquitoes during log phase of growth (Relative percentage)

Fatty acid	Medium	Aedes aegypti[a]	Aedes albopictus[a]	Culex quinque-fasciatus[b]	Culex tritae-niorhynchus[b]
14:0[c]	1.9	2.0	2.6	1.5	1.3
15:0	0.6	1.4	1.5	-	tr
16:0	21.7	17.4	21.2	14.4	16.6
16:1	5.9	15.6	14.5	11.0	8.2
18:0	11.5	7.0	7.8	8.1	12.0
18:1	30.7	38.2	38.2	37.9	39.3
18:2	13.6	6.6	5.1	9.2	9.6
18:3	1.4	0.9	0.5	1.6	1.3
20:0	tr	3.9	3.1	5.2	3.4
20:3	1.7	1.0	0.9	3.2	0.9
20:5	tr	1.1	1.4	0.8	0.8

a From McMeans et al., 1975

b From McMeans et al., 1976

c Number of carbon atoms in acid: number of double bonds

tr = trace < 0.5%

contrast, 67.3% and 76.0% of the neutral lipids in A. aegypti and A. albopictus, respectively, occurred as triglycerides compared to only 19.2% in the medium. This corresponds to the findings in insects that the fatty acids are transported and stored as the triglycerides. Much of this conversion is believed to occur in the fat body of the insect (Gilmor, 1965). However, as far as can be determined the cells of the lines used by Yang and his co-workers did not originate from fat body tissue. As the cells entered the stationary phase of the growth cycle the nature and amounts of lipid present changed. In both cell lines there was a decrease in the level of triglycerides, 10% in the A. aegypti line and 30% in the A. albopictus cells. There also was a slight shift in the fatty acid composition of the triglycerides as the cells entered the stationary phase. In both Aedes cells there was an increase in the degree of unsaturation of the constituent fatty acids. A. albopictus cells contained fatty acids with increasingly higher carbon numbers in the triglycerides but in A. aegypti the average carbon number of the fatty acids decreased.

The predominent fatty acid in the phospholipids in both cell lines was oleic, 38.1 and 38.6% respectively (McMeans et al., 1976). This was somewhat higher than the level of oleic acid in the medium, 30.4%. Palmitic was the second highest, 15.8 and 17.1%, respectively. These levels were lower than that in the medium, 20.7%. The phospholipids contained less oleic acid and more linoleic acid in the stationary phase than in the growth phase.

Similar studies have been reported by Jenkin and Townsend and their coworkers on cells from the eucalyptus moth, Antheraea eucalypti (Jenkin et al., 1971; Townsend et al., 1972). These studies have caused some confusion in the literature because the cell line was believed to be A. aegypti at the time and was so described in the two papers. Later it was properly identified as the Antheraea cell line and so the data should be considered as from a species of Lepidoptera. These studies were done in a medium supplemented with fetal bovine serum rather than the newborn calf serum used in the medium for the mosquito cells and this resulted in a slightly different distribution of fatty acids in the medium. The most prevalent acid was linoleic acid, 42.9% of total fatty acids. This was almost 3 times the percent of the total fatty acids represented by linoleic acid in newborn serum. The percent of the total neutral lipid in the form of free fatty acids was 8.6% in the fetal serum as opposed to 15.1% in the newborn serum. The triglycerides constituted only 4.6% of the neutral lipids in fetal bovine serum but were 19.2% of the newborn serum.

The most noticeable difference in the lipid composition of the cell of the two lines was that the moth cells contained relatively higher levels of free fatty acids, 49.5% of the neutral lipid, as opposed to 5.0% or less in the mosquito cells. In contrast, only 20.5% of the neutral lipids in moth cells were in the form of triglycerides compared to 67% or more in the Aedes cells. The most prevalent fatty acid in the cells of both moths and mosquitoes, as determined from the total lipids, was oleic acid. This was true despite the high percentage of linoleic acid in the fetal bovine serum.

These studies provided good evidence that insect cells derived from tissues other than fat body were capable of considerable alteration of the fatty acids present in the growth medium provided. Certainly, chain elongation and alteration of the level of saturation were possible. The studies did not provide any evidence of a nutritional requirement for individual fatty acids. McMeans et al. (1975) suggest that Aedes cells may not need any preformed lipids in the medium. Louloudes et al. (1973) studied the depletion of fatty acids from cell culture medium by fall armyworm cells to determine if any of the fatty acids provided by the serum supplements were completely utilized and therefore might be growth limiting. They were not able to detect the complete disappearance of any fatty acid initially present in significant amounts. The medium was supplemented with fetal bovine serum, bovine serum albumin and hemolymph from the silkworm, Bombyx mori L. and contained 40.93 μg of fatty acid/ml initially. After three growth cycles the concentration had been reduced to 19.66 μg/ml. Seven fatty acids, palmitic (16.0), palmitoleic (16:1), stearic (18:0), oleic (18:1), linoleic (18:2), linolenic (18:3), and arachidonic (20:4) made up approximately 92% of the total mass. The changes in the levels of these seven fatty acids during the study are shown in Table 2. In the first passage the medium supported normal cell growth, approximately a 10 fold increase in cell number. No fatty acids were completely depleted although the level of arachidonic acid decreased 63%, linolenic acid decreased 58%, and linoleic decreased 29%. It would appear that the cells are utilizing these fatty acids directly from the medium; however, the increases in the levels of oleic, stearic and palmitoleic could indicate that the decreases occurred because the fatty acids were simply altered by the cells and then excreted rather than incorporated into the cell. This was the most likely process because examination of the level of these fatty acids in the cells revealed no increase. Cell growth in the second and third passage was markedly reduced, by two-thirds in the second passage and by 95% in the third. The cells from these passages had increased levels of fatty acids as is typical of many animal cells in culture when growth stops (King et al., 1959; Bensch et al., 1961).

The addition of various materials rich in linolenic and linoleic acids was shown by Vaughn and Louloudes (1972) to increase the cell yield of serum supplemented medium. Oils such as linseed (linolenic acid) or safflower (linoleic acid) increased the cell yield 40%, whereas menhaden oil (palmitic acid) even at twice the level of safflower oil or four times the level of linseed oil, gave no increase in cell yield. The triglycerides, trilinolenin, and trilinolein gave similar increases in cell yield. However, it required ten times more trilinolein to achieve equivalent cell yields.

Thus, although a specific requirement for fatty acids has not been demonstrated, the addition of fatty acids with two or more double bonds is stimulatory even in the presence of animal sera. Recently, a chemically defined culture medium was reported which supported serial passage of insect cells (Wilkie et al., 1980). Such a medium, of course, represents the ideal for the study of cell nutrition. This medium contains the following fatty acids: stearic, myristic, oleic, linoleic, linolenic, palmitic, palmitoleic, and arachidonic as well as the triglycerides trilinolein and trilinolenin. A S. frugiperda cell line has been cultured for 20 passages, equivalent to approximately 60 population doublings in this medium. This growth is certainly sufficient to deplete

TABLE 2.

Depletion of several fatty acids by repeated use of cell culture medium[a]

Fatty acid	Preincubation	First	Second	Third
Palmitic	20.93	20.71	19.90	20.15
Palmitoleic	3.79	7.14	8.50	9.65
Stearic	12.92	16.55	12.35	12.28
Oleic	17.66	28.39	24.55	25.55
Linoleic	10.82	7.70	8.57	8.52
Linolenic	22.98	9.59	13.84	10.16
Arachidonic[b]	4.19	1.54	2.47	0.76

[a] Data from Louloudes et al., 1973
[b] Fatty acid was trapped individually and analyzed by gas-liquid chromatographic mass spectrometer.

any fatty acids carried over from the serum medium but the value of the individual fatty acids has not yet been reported. The authors also reported success in culturing two mosquito cell lines from A. aegypti and Anopheles gambiae in the medium for five passages after modification of the salt composition and glucose content. Again requirements for individual fatty acids have not yet been reported but the system has great potential for the determination of the fatty acid requirements and their metabolism by insect cells.

STEROLS

All animal and some plants require sterols for two purposes: One as an indispensable component of the membranes of cellular and subcellular structures, and two, as precursors for essential steroid hormones (Svoboda et al., 1978). In insect cell cultures the sterols are required only as components of the cellular membranes since none of the established cell lines are known to synthesize hormones. As is true of the insect, the necessary sterols must be obtained from exogenous sources as the cells are incapable of sterol synthesis. For all established insect cell lines the sterol utilized is cholesterol and the source is the various animal sera added as medium supplements. The utilization of the serum sterols was first demonstrated by Vaughn et al. (1971) with the Grace Antheraea cells. Their medium, supplemented with fetal bovine serum, bovine serum albumin and chicken egg ultrafiltrate contained 20.2 μg of sterol per milliliter. They found that the cells incorporated only 4.01% of the medium sterols during growth. However, this accounted for only 11.22% of the original sterol. When the medium was further fortified with ^{14}C-cholesterol bound to the fetal bovine serum, 10% of the recovered ^{14}C-cholesterol was found in the cells but of the label remaining in the medium only 12.5% was in cholesterol; the remainder was in more polar compounds indicating that considerable alteration of the molecule had occurred but the nature of the alterations were not known.

In a later study using the same cell line but with the medium supplemented only with insect hemolymph and bovine serum albumin Gilby and McKellar (1974) quantitatively determined the uptake of medium sterols. Analysis of their medium indicated that the sterol component consisted of cholesterol and β-sitosterol derived entirely from the hemolymph component. They determined that about 67% of the cholesterol and 85% of the sitosterol were lost from the medium during incubation without cells. If the level in the cultured medium was corrected for this loss, then the levels incorporated into the cells approximately equalled the amounts which disappeared from the medium. They concluded that the conservation of the sterols indicated that their major role was as components of the cellular membranes.

The studies of Gilby and McKellar indicated that the Antheraea cells utilized the media sterols as they occurred and did not convert the sitosterol to cholesterol before incorporating it into cell structures. Louloudes (1976) incubated cells derived from the fall armyworm, Spodoptera frugiperda, with ^{3}H-sitosterol and determined that 93% of the radioactivity recovered from the cells was sitosterol. Thus, cells of this line did not convert the sitosterol to cholesterol. Neither could these cells convert desmosterol to other sterols. However, when excised midgut tissue from fall armyworm larvae was incubated with ^{14}C-desmosterol 6% of the tissue sterol was cholesterol, indicating that this tissue had some ability to convert the dietary sterols.

Two studies have been reported which demonstrate the inability of insect cells to synthesize sterols from a simple precursor such as acetate. In the study by Louloudes (1976), referred to above, ^{14}C-acetate (8 x 10^6 CPM) was incorporated into the medium used to culture cells from S. frugiperda. After 48 hours incubation the sterols were isolated from the cells as their digitonides and examined for radioactivity. No radioactivity was detected by the third sequential digitonide precipitation of the purified sterols, thus indicating that the cells were incapable of de novo sterol biosynthesis. This confirmed an earlier study of Cohen and Gilbert (1975) who used the TN-368 cell line derived by Hink from the cabbage looper, Trichoplusia ni. In their study ^{14}C-acetate was incorporated in Grace's medium devoid of lipids and after 5 hours the cellular lipids were extracted and chromatographed. Although the lipid fraction exhibited considerable radioactivity, careful analysis revealed that the label was in long chain alcohols and that acetate was not incorporated into sterol in T. ni cells.

Thus it appears that insect cells in culture like the animals from which they were derived cannot synthesize the required sterols from simple substrates. To our knowledge, just where the pathway is blocked has not been determined. In addition it would appear that the cells in the established cell lines are incapable of some of the sterol interconversions which the whole animal can make. This is quite likely a reflection of the limited types of tissues that have been used to initiate cell cultures; as cultures are prepared from a wider variety of tissues, more may be learned about sterol metabolism.

These results indicate that sterol in some form may be an essential nutrient for cells in culture. This requirement was definitely established by Brooks et al. (1980) who demonstrated that cholesterol was essential for the growth of cell lines from the cockroach, Blattella germanica. Medium containing cholesterol at a final concentration of 22 μg/ml supported growth of two of the four cell lines tested for 22 to 39 subcultures as of the time the report was prepared whereas the cells did not grow in the absence of sterol. No attempt to determine the optimum concentration was reported.

The method of preparing the stock solutions of cholesterol was of critical importance for the successful long term culture of the cells. The dissolution of the cholesterol in boiling ethanol, followed by dilution with distilled water and further boiling to reduce the final volume by nearly one-half was the most successful. The use of Tween 80 as an emulsifier was not successful as the Tween 80 was toxic in the absence of serum. If cholesterol and lecithin were sonicated together in water to produce a fine colloidal dispersion a particulate material resulted which eventually settled out. This supplement, including the fine particulate material, would not support continued cell growth. The use of cold ethanol to dissolve the cholesterol also was an unsatisfactory method of preparing a cholesterol stock solution.

The authors' tentative conclusion was that some undefined physico-chemical reaction between the ethanol and the cholesterol occurred which made the cholesterol utilizable. Whatever the change was, the experience of Brooks and her co-workers illustrates some of the problems in studying lipids in cell culture and seems to emphasize the point that the lipid materials must be presented to the cells in a useable form before a true evauation of their importance can be made.

GLYCERIDES AND PHOSPHOLIPIDS

The free fatty acid content of insect tissues is relatively low, as the di- or triacylglycerols or phosphoglycerides predominate. The fat body contains high levels of triacylglycerol and fatty acids in the hemolymph are transported as diglycerides attached to carrier proteins. In some insects glycerides serve as energy sources during strenuous activity (Downer, 1978). Since none of these specialized tissues have been successfully used as sources of cells for established cell lines the data obtained from studies with such cell lines is difficult to relate to metabolism in the whole insect.

The glyceride composition of the cells from several insect cell lines is shown in Table 3. The neutral lipid content of the four mosquito cell lines shown in the table demonstrates the characteristic high triglyceride

TABLE 3.

The glyceride composition of the cultured cells from several insect species

Glyceride	Antheraea eucalypti[a]	Aedes aegypti[b]	Aedes albopictus[b]	Culex quinque-fasciatus[c]	Culex tritae-niorhynchus[c]
Monoglyceride	13.6[d]	trace	0.2	0.8	0.5
Diglyceride	3.2	2.7	1.9	2.2	1.0
Triglyceride	20.5	67.3	76.0	80.4	80.2

[a] From Townsend et al., 1972
[b] From Yang et al., 1976a
[c] From Yang et al., 1976b
[d] Percent of neutral lipids

levels associated with insect tissue. The Antheraea cells analyzed by Townsend et al. (1972) had nearly equal levels of triglycerides and monoglycerides, 20% and 13% respectively. In the mosquito cells the triglycerides predominated, accounting for 67-80% of the total glyceride composition of the cells. Moreover, the media in which these mosquito cells were cultivated contained much lower levels of glycerides, 19.7% (Yang et al., 1976a). Thus it would appear that much of the free fatty acid content of the media was converted by the mosquito cells into glycerides.

Some indication that glycerides play a role in the nutrition of insect cells in culture is shown by the work of Goodwin and Adams (1980). They reported that in serum-free medium the addition of glycerol supported cell growth whereas none occurred in its absence. Their studies showed that a concentration of 2g/l was optimum. The addition of α-glycerolphosphate further improved cell growth. No free fatty acids or other glycerides were added to the medium.

The role of the phospholipids in the nutrition and metabolism of cultured insect cells is poorly understood. In insects choline can serve as an adequate substrate for the synthesis of phosphotidylcholine (Downer, 1978). It was incorporated into insect cell culture medium first by Grace (1958) at a concentration of 100 µg/ml. Later, Nagle (1969) established a concentration of 50 mg/ml as optimum for A. aegypti cells. (It should be noted that these cells were quite likely A. eucalypti although Nagle did not give the source). Goodwin et al. (1978) replaced choline with acetyl-ß-methyl-choline Cl but did not define the relative value of the two compounds in promoting cell growth. There have been no studies which indicated that choline was incorporated into phospholipids as the result of cell synthesis. However, Cohen and Gilbert (1975) demonstrated that cultured cells from the cabbage looper (TN-368 cell line) were able to synthesize phospholipids from acetate. The cells were washed free of medium containing serum and other undefined materials and incubated in Grace's medium without supplements but containing 7×10^6 DPM [1-^{14}C] sodium acetate for 5 hours. The phospholipids were separated by chromatography and found to contain 97.6% of the total added activity. Phosphatidylethanolamine constituted 38.0% and phosphatidycholine 33.7% of the total phospholipids.

These levels correspond closely to the levels found by Jenkin and his co-workers (1971) in the cells cultured from A. eucalypti. In these moth cells phosphotidylethanolamine accounted for 31.1% of the total phospholipid and phosphotidylcholine 42.0%. Other phospholipids identified were: sphingomyelin 15.2%, lysophosphatidylcholine 7.7%, and phosphotidylinositol 4.0%. These cells were cultured in medium supplemented with fetal bovine serum whose principal phospholipids were reported to be phosphotidylcholine and sphingomyelin with only 4.7% phosphatidylethanolamine.

In contrast, in Aedes albopictus cells cultured in medium supplemented with the same serum phosphotidylethanolamine constituted 46.4% of the total phospholipids and phosphotidylcholine only 22.2% (Luukkonen et al., 1973). Jenkin and his co-workers (Jenkin et al., 1975) analyzed the phospholipids in the same cell line and in a cell line from A. aegypti and reported that in A. albopictus phosphatidylethanolamine constituted 53.6% of the total phospholipid and phosphotidylcholine 30.7%. In the cultured A. aegypti cells phosphatidylethanolamine and phosphotidylcholine constituted 48.8 and 37.4 percent, respectively, of the total phospholipid in the cells. Their medium was supplemented with newborn calf serum which has a very high percentage of the phospholipid content (71.8%) in the form of phosphotidylcholine. Further studies with cultured cells from C. tritaeniorhynchus gave similar results, 45.1% of the total phospholipids were phosphotidylethanolamine and 37.2% phosphotidylcholine (Jenkin et al., 1976). However, the cells from C. quinquefasciatus had nearly equal amounts of the two phospholipids but slightly less phosphatidylethanolamine (41.3%) than phosphotidylcholine (43.0%).

The use of bovine sera as a media supplement results in a high percentage of phosphotidylcholine in the media. It is clear from the results of these studies that the cultured cells must either convert this phospholipid to phosphatidylethanolamine or else synthesize this compound from substrates in the media as Cohen and Gilbert (1975) demonstrated their cell lines were capable of doing. This cannot be determined from the studies reported above. The major site of this synthesis in the whole insect is reported to be in the fat body (Downer, 1978). However, since these cultures are not known to contain fat body cells it would appear that other cells in the insect are also capable of such synthesis.

It should also be noted that the cells derived from both Lepidoptera and Diptera contained proportions of phosphatidylethanolamine and phosphotidylcholine corresponding to those proportions found in the whole insect. That is, the cells from Lepidoptera had proportionately more phosphotidylcholine and those from Diptera had more phosphatidylethanolamine. An exception was the C. quinquefasciatus cells in which the two phospholipids occurred in essentially equal amounts.

From these studies it appears unlikely that cultured insect cells have any requirements for exogenous sources of specific types of phospholipids but that all the requirements can be met either by synthesis from simple

precursors or by modification of the exogenous phospholipids. Although it has been established that there are
differences in the phospholipid composition of cells from
different orders or genera of insects, no specific cell
functions have yet been identified with these differences. Cell culture might be an appropriate tool for in-
vestigating the role of phospholipids in such phenomena
as receptor site specificity and membrane transport.

HORMONES

One of the problems faced by the first workers who
tried to culture insect cells was whether or not the lack
of success in establishing cell cultures was due to an
absence or an improper balance of growth stimulating hormones. It was soon established that explanted tissues
would respond to exposure to hormones and that in some
systems the in vivo functions of tissues could be dupli-
cated in vitro if the explants were given the correct
hormone treatment. This research has been recently reviewed by Marks (1976, 1980) and interested readers are
referred to those articles. Only the effects of hormones
on established cell cultures will be discussed here.
When the early attempts to culture insect cells were
unsuccessful it was considered possible that some of the
steroid hormones would be required in the media in order
to obtain cell division and continued growth. However,
the culture of insect cells in media devoid of any insect
materials which might contain hormones indicated that
this was not true (Yunker et al., 1967; Singh, 1967).
Then Mitsuhashi and Grace (1970) tested the effects of
20-hydroxyecdysone, ponasterone A, proteinaceous brain
hormone and farnesol on the growth of cells of the Grace
Antheraea line. The brain hormone and farnesol had no
stimulating effect on the cells but 20-hydroxyecdysone
and ponasterone A stimulated growth at concentrations of
0.1 µg per ml but the effect was not marked and did not
occur at all in media which did not contain insect hemolymph. The authors concluded that the insect hormones
were not growth promoting factors for insect cells cultivated in vitro. Farnesol did not effect cell growth in
the presence of serum but at 10.0 µg per ml in protein-
free medium inhibited cell growth. Combinations of farnesol and 20-hydroxyecdysone also were not growth stimulating.
The effect of protein on the cell-hormone interaction
was also noted by Courgeon (1972) in studies done with
the Kc line of cells from Drosophila melanogaster. In
these studies cultures treated with 0.3 to 3 µg/ml of
ecdysone grew significantly higher numbers of cells than
control cultures. However, when the calf serum was omit-
ted from medium no growth stimulation was obtained at any

hormone concentration. The relative increase in the number of cells was greater when the ecdysone was added to medium with low levels of serum but even in medium containing 20% fetal calf serum ecdysone was capable of stimulating increased growth. Inokosterone and 20-hydroxyecdysone did not stimulate cell growth even in the presence of protein. Cell growth was impaired by 20-hydroxyecdysone at a concentration as low as 0.006 μg/ml. At this concentration the cells elongated to take a fibroblast-like morphology and after this change no longer multiplied.

No such differences in the action of ecdysone and 20-hydroxyecdysone were detected by Wyss (1976) using a clone of hypotetraploid cells isolated from the Kc line. Neither was he able to demonstrate a dependence on serum proteins for ecdysone to be effective. Wyss found that the optimum concentrations for stimulation of cell growth were 3 ng/ml for 20-hydroxyecdysone and 300 ng/ml for ecdysone whereas higher concentrations inhibited cell growth. Both effects were inhibited by ethyl dichloro-farnesoate, a juvenile hormone analog.

The cytological effects of 20-hydroxyecdysone on cells of the Kc Drosophila line were further studied by Rossett (1978). Cells treated with 10 ng/ml 20-hydroxyecdysone became spindle-shaped and began to orientate 24 hr after exposure to the hormone. Twelve hours later the cells aggregated in clumps and within 2 days most of the cells had gathered in large aggregates. These changes are in agreement with those described by Courgeon earlier (1972). Biochemically, the cells stop synthesizing DNA within 10 hrs after treatment; RNA synthesis progressively declines and eventually protein synthesis stops, as indicated by incorporation of ^{14}C-tyrosine.

Lanir and Cohen (1978) reported that treatment of cells of the Kc line with 1 μg/ml concentrations of 20-hydroxyecdysone resulted in irreversible morphological changes that were very similar to those described by Courgeon and by Rosset. In addition time lapse cinematography demonstrated that the hormone treatment stimulated cell migration and cell membrane movements. These effects were not produced by treatment of the cells with 22-iso-ecdysone.

Although the Drosophila cell lines have been the most widely used to study the action of the ecydsteroids on cultured cells they are not the only ones used. Cohen and Gilbert (1972) used two cell lines from Lepidoptera in their studies. They found that cells of the A. eucalypti and the T. ni lines did not respond to either ecdysone or 20-hydroxycdysone at concentrations of 1-50 μg/ml.

Studies with tritium labeled ecdysone revealed essentially no uptake of ^{3}H-ecdysone and a very limited uptake of 20-hydroxyecdysone after 48 hr. The addition of serum proteins from hemolymph or bovine serum improved the uptake of 20-hydroxyecdysone only very slightly. They found that juvenile hormone on the other hand was a potent inhibitor of both RNA and protein synthesis in cells of both lines and caused cell swelling and plasma membrane lesions.

Marks and Holman (1979) found that cells of lines derived from another lepidopteran, Manduca sexta, showed a marked response to treatment with 20-hydroxyecdysone. Treated cells became attenuated and the apparent diameter of the nucleus decreased by nearly 50%; blebbing of the cytoplasm occurred and cytoplasmic debris appeared in the medium. In addition, the number of dead cells increased sharply. They reported that the threshold level necessary to induce the response was between 6 and 7 x 10^{-5} M with a maximum response at about 10^{-6} M, which is similar to the levels found to produce the changes in the Drosophila Kc cells. Also the timing of the appearance of the various responses was very similar. Unlike the Kc cells, these changes were not irreversible and when subcultured in the absence of ecdysone the cells continued to live and multiply. Continued culture in the presence of ecdysone eventually led to the selection of a strain of the Manduca cells that would no longer respond to 20-hydroxyecdysone and another strain which responded to 20-hydroxyecdysone but not the ecdysone. The parent cells responded to ecdysone only after converting it to 20-hydroxyecdysone so the conclusion was that the new strain had lost the ability to perform this hydroxylation.

Recent studies have attempted to establish whether or not these responses represent normal responses and whether or not the permanent cell lines could be used as model systems for studying cell-hormone interactions. Most of the cell lines used in these studies have been derived from whole embryos or from complex organs so that the tissue of origin is uncertain and the parental type within an insect is unknown. Therefore, there are no specific cellular responses from the insect which can be directly related to the responses observed in cell culture. Solution of this problem has been approached in several ways; the search for steroid hormone receptors, the effects on cells of various related compounds already tested in insects, the genetic mechanisms controlling the sensitivity to hormones, and the biochemical events which can be influenced in the cell by treatment with ecdysteroids. These studies were reviewed in some detail by Cherbas et al. (1980), Berger et al. (1980), and Wyss (1980) in the proceedings of a recent conference on invertebrate cell culture and a summary of their papers is given here.

Cherbas and his co-workers (1980) have tested 60 compounds for their ability to elicit the morphological response and the appearance of acetyl-cholinesterase. Fourteen compounds elicited both responses in Kc-H cells. Of the 46 compounds which did not produce a response, only six were known to have ecdysteroid activity and based on their chemical structure they may be converted in vivo to active compounds. After examining the binding of ecdysone and 20-hydroxyecdysone they concluded that both compounds act by binding to the same receptors. The receptor sites on the cultured cells had an affinity and specificity similar to sites of the receptor system in intact Drosophila tissues. No evidence for other receptors was found, and therefore, they believe that the Kc cells are a valid model system for studying the mechanism by which hormones control gene activity.

The genetic control of the sensitivity and resistance to ecdysone in clones obtained from permanent cell lines has also been studied in some detail (see review by Wyss, 1980). Resistant cells have been found to occur spontaneously in the absence of ecdysone and this characteristic is stable in these cells even for extended periods of culture in the absence of the hormone. Thus the characteristic appears to be a true mutation rather than an inducible function.

One of the difficulties in relating in vitro studies to in vivo phenomena is the uncertainty of the origin of the cells in the permanent cell lines. Wyss (1980) has attempted to overcome this problem by using cell hybrids, newly isolated cells fused to cells from permanent cell lines. He reports that cells from 6-hour-old embryos fused to ecdysone-sensitive MDR3 cells resulted in resistant hybrid cells. However, cells from dissociated eye-antennal discs of wandering stage third instar larva when fused to the MDR3 cells produced sensitive hybrids. Thus, this cell fusion method may provide a very sensitive probe with which to study the tissue specific aspects of the hormone response.

These studies dealt only with the growth inhibitory effect of ecdysone, other effects such as the change in morphology or in biochemical activity are not necessarily related and must be investigated independently. After studying four cell lines from Drosophila (Kc, L1, L2 & L3) Berger and his colleagues (1980) reached the following conclusions: First, the synthesis of only a small number of proteins are ever affected by ecdysone. Second, most of the changes that occurred following exposure to ecdysone were quantitative in nature. Third, to a degree, each cell line responded uniquely to hormone exposure. Fourth, the six proteins whose synthesis increased substantially in all of the cell lines included two cytoplasmic actins and at least one of the tubulins. These increases may account for the increased cell mobil-

ity and change in cell shape noted as a typical response
to hormone exposure. However, they concluded that there
was little evidence that the changes they observed in
these cells had in vivo counterparts, nor was there sufficient evidence to establish whether or not the responses observed in cultured cells actually were primary effects of ecdysone. Thus, they concluded that cells from
permanent cell lines could not yet be said to be a valid
model for studying the primary effects of ecdysone.

CONCLUSIONS

Little attention has been given to the nutritional
requirements of cultured insect cells for lipids with the
exception of the sterols. It has been demonstrated that
cultured cells cannot synthesize sterols and from this
it has been assumed that there is a requirement for exogenous sterol but an absolute requirement has so far only
been demonstrated for cells from the cockroach (Brooks
et al., 1980). Similarly, it has been clearly demonstrated that fatty acids added to the culture media will result in increased cell yield. However, no individual
fatty acids have been established as essential nutrients.

The need to supplement the existing cell culture
media with sera, either invertebrate or vertebrate, has
obscured any lipid requirement because these materials
are naturally high in lipids. The recent development of
serum-free and chemically defined media not only provides
the means for studying lipid requirements but makes it
necessary in order to duplicate the results previously
obtained with the serum-supplemented media.

Continuously cultured insect cells are potentially
very useful in the study of the metabolism of lipids.
However, as has been pointed out in the course of this
review there are several limitations which must be understood when planning studies or interpreting data. The
most important is that the tissue of origin for most cell
lines is unknown with the exception of those lines derived from hemocytes. The other limitation is that the
behavior of cells in the tissue of a whole insect can be
modified by other cell types present in the tissue or the
organ. Thus, even if the parent cell type of the cultured cells is known the effects of other cell types are
absent in culture.

If these limitations are kept in mind, then studies
on lipid metabolism in cultured cells will have considerable value in clarifying our understanding of the processes as they occur in vivo. As the understanding of cell
nutrition increases and the differentiated cells of various tissues and organs can be maintained and then cultured for extended periods of time, the value of culture
systems for metabolic studies will increase considerably.

REFERENCES

Bensch, K.G., D.W. King, and E.L. Socolow. (1961) The source of lipid accumulation in L cells. J. Biophys. Biochem. Cytol. 9: 135-139.

Berger, E.M., M. Frank, and M.C. Abell. (1980) Ecdysone induced changes in protein synthesis in embryonic Drosophila cells in culture. In "Invertebrate Systems In Vitro", E. Kurstak, K. Maramorosch and A. Dübendorfer, eds. Elsevier/North Holland Biomed. Press, Amsterdam. 195-208.

Brooks, M. A., K. R. Tsang, and F. A. Freeman. (1980) Cholesterol as a growth factor for insect cell lines. In "Invertebrate Systems In Vitro", E. Kurstak, K. Maramorosch and A. Dübendorfer, eds. Elsevier/North Holland Biomed. Press, Amsterdam. 66-77.

Cherbas, L., P. Cherbas, C. Savakis, G. Demetri, M. Manteuffel-Cymborowska, C.D. Yonge, and C.M. Williams. (1980) Studies of ecdysteroid action on a Drosophila cell line. In "Invertebrate Systems In Vitro", E. Kurstak, K. Maramorosch and A. Dübendorfer, eds. Elsevier/North Holland Biomed. Press, Amsterdam. 218-228.

Chippendale, G. M. (1972) Insect metabolism of dietary sterols and essential fatty acids. In "Insect and mite Nutrition Significance and Implications in Ecology and Pest Management", J.G. Rodriguez, ed. Elsevier/North Holland Biomed. Press, Amsterdam. 423-435.

Cohen, E. and L.I. Gilbert. (1972) Metabolic and hormonal studies on two insect cell lines. J. Insect Physiol. 18: 1061-1076.

Cohen, E. and L.I. Gilbert. (1975) Lipid synthesis and RNA thermolability in the Trichoplusia ni cell line. Insect Biochem. 5: 671-677.

Courgeon, A-M. (1972) Effects of α- and ß-ecdysone on in vitro diploid cell multiplication in Drosophila melanogaster. Nature New Biol. 238: 250-251.

Downer, R.G.H. (1978) Functional role of lipids in insects. In "Biochemistry of Insects", M. Rockstein, ed. Academic Press, New York. 57-92.

Gilby, A.R. and J.W. McKellar. (1974) The utilization of sterols and other lipids by insect cells grown in vitro. J. Insect Physiol. 20: 2219-2224.

Gilmor, D. (1965) "The Metabolism of Insects." W.H. Freeman, San Francisco. 195 pp.

Goodwin, R.H. and J.R. Adams. (1980) Nutrient factors influencing viral replication in serum-free insect cell line culture. In "Invertebrate Systems In Vitro", E. Kurstak, K. Maramorosch and A. Dübendorfer, eds. Elsevier/North Holland Biomed. Press, Amsterdam. 493-509.

242

Goodwin, R.H., G.J. Tompkins, and P. McCawley. (1978) Gypsy moth cell lins divergent in viral susceptibility. I. Culture and identification. In Vitro 14: 485-494.

Grace, T.D.C. (1958) Effects of various substances on growth of silkworm tissues in vitro. Aust. J. Biol. Sci. 11: 407-417.

Jenkin, H.M., E. McMeans, L.E. Anderson, and T-K. Yang. (1975) Comparison of phospholipid composition of Aedes aegypti and Aedes albopictus cells obtained from logarithmic and stationary phases of growth. Lipids 10: 686-694.

Jenkin, H.M., E. McMeans, L.E. Anderson, and T-K. Yang. (1976) Phospholipid composition of Culex quinquefasciatus and Culex tritaeniorhynchus cells in logarithmic and stationary growth phases. Lipids 11: 697-704.

Jenkin, H., D. Townsend, S. Makino, and T-K. Yang. (1971) Comparative lipid analysis of Aedes aegypti and monkey kidney cells (MK-2) cultivated in vitro. Curr. Topics Microbiol. Immunol. 55: 92-97.

King, D.W., E.L. Socolow, and K.G. Bensch. (1959) The relationship between protein synthesis and lipid accumulation of L strain cells and Ehrlich ascites cells. J. Biophys. Biochem. Cytol. 5: 421-431.

Lanir, N. and E. Cohen. (1978) Studies on the effect of the moulting hormone in a mosquito cell line. J. Insect Physiol. 24: 613-621.

Louloudes, S.J. (1976) Sterol metabolism in the insect cell line IPLB-SF-21B (Spodoptera frugiperda). Proc. 1st Internat. Colloq. Invertebr. Pathol. IXth Ann. Meeting for Invertebr. Pathol. Aug. 29-Sept. 3, 1976. Queen's Univ., Kingston, Ont., Queen's Univ. Printing Dept., Kingston. 399-400.

Louloudes, S. J., J. L. Vaughn, and K. Dougherty. (1973) Fatty acid profiles of cells from the insect line IPRL-21 (Spodoptera frugiperda) and of the tissue culture medium after repeated use. In Vitro 8: 473-479.

Luukkonen, A., M. Brummer-Korvenkontio, and O. Renkonen. (1973) Lipids of cultured mosquito cells (Aedes albopictus) comparison with cultured mammalian fibroblasts (BHK 21 cells). Biochim. Biophys. Acta 326: 256-261.

Marks, E.P. (1976) The uses of cell and organ cultures in insect endocrinology. In "Invertebrate Tissue Culture Research Applications", K. Maramorosch, ed., Academic Press, New York. 117-132.

Marks, E.P. (1980) Insect tissue culture: an overview, 1971-1978. Ann. Rev. Entomol. 25: 73-101.

Marks, E.P. and M. Holman. (1979) Ecdysone action on insect cell lines. In Vitro 15: 300-307.

McMeans, E., T-K. Yang, L.E. Anderson, and H.M. Jenkin. (1975) Comparison of lipid composition of Aedes aegypti and Aedes albopictus cells obtained from logarithmic and stationary phases of growth. Lipids 10: 99-104.

McMeans, E., T-K. Yang, L.E. Anderson, S. Louloudes, and H.M. Jenkin. (1976) Comparison of the lipid composition of Culex quinquefasciatus and Culex tritaeniorhynchus cells obtained from the logarithmic and stationary phases of growth. Lipids 11: 28-33.

Mitsuhashi, J. and T.D.C. Grace. (1970) The effects of insect hormones on the multiplication rates of cultured insect cells in vitro. Appl. Ent. Zool. 5: 182-188.

Nagle, S.C. Jr. (1969) Improved growth of mammalian and insect cells in media containing increased levels of choline. Appl. Microbiol. 17: 318-319.

Paphadjopoulos, D., T. Wilson, and R. Taber. (1980) Liposomes as vehicles for cellular incorporation of biologically active macromolecules. In Vitro 16: 49-54.

Rosset, R. (1978) Effects of ecdysone on a Drosophila cell line. Exp. Cell Res. 111: 31-36.

Rothblat, G. H., L. Y. Arbogart, L. Ouellette, and B. V. Howard. (1976) Preparation of delipidized serum protein for use in cell culture systems. In Vitro 12: 554-557.

Singh, K.R.P. (1967) Cell cultures derived from larvae of Aedes albopictus (Skuse) and Aedes aegypti (L.) Curr. Sci. (India) 36: 506-508.

Svoboda, J.A., M.J. Thompson, W.E. Robbins, and J.N. Kaplanis. (1978) Insect steroid metabolism. Lipids 13: 742-753.

Townsend, D., H.M. Jenkin, and T-K. Yang. (1972) Lipid analysis of Aedes aegypti cells cultivated in vitro. Biochem. Biophys. Acta. 260: 20-25.

Vaughn, J.L. (1971) Design of cell culture media. In "Invertebrate Tissue Culture," C. Vago, ed. Academic Press, New York, Vol. 1. pp. 3-40.

Vaughn, J.L. and S.J. Louloudes. (1972) The role of sterols and fatty acids in the nutrition of insect cells in culture. In "Insect and Mite Nutrition Significance and Implications in Ecology and Pest Management", J.G. Rodriguez, ed. Elsevier/North Holland, Amsterdam. 376-385.

Vaughn, J.L., S.J. Louloudes, and K. Dougherty. (1971) The uptake of free and serum-bound sterols by insect cells in vitro. Curr. Topics Microbiol. Immunol. 55: 92-97.

Wilkie, G.E.I., H. Stockdale, and S.V. Pirt. (1980) Chemically-defined media for production of insect cells and viruses in vitro. Proc. 3rd Gen. Meeting

244

Europ. Soc. Animal Cell Technol. Oct. 2-5, 1979. Keble College, Oxford, U.K. Devel. Biol. Standardization 46: 29-37. S. Karger, London.

Wyss, C. (1976) Juvenile hormone analogue counteracts growth stimulation and inhibition by ecdysones in clonal Drosophila cell line. Experientia 32: 1272-1274.

Wyss, C. (1980) Cell hybrid analysis of ecdysone sensitivity and resistance in Drosophila cell lines. In "Invertebrate Systems In Vitro", E. Kurstak, K. Maramorosch, and A. Dübendorfer, eds. Elsevier/North Holland Biomed. Press, Amsterdam. 279-289.

Yang, T-K., E. McMeans, L.E. Anderson, and H.M. Jenkin. (1976a) Neutral lipids of Aedes aegypti and Aedes albopictus cells cultured in vitro. J. Invertebr. Pathol. 27: 161-169.

Yang, T-K., E. McMeans, L.E. Anderson, and H.M. Jenkin. (1976b) Neutral lipid composition of Culex quinquefasciatus and Culex tritaeniorhynchus cells at two phases of growth. Lipids. 11: 21-27.

Yunker, C.E., J.L. Vaughn, and J. Cory. (1967) Adaptation of an insect cell line (Grace's Antheraea cells) to medium free of insect hemolymph. Science 155: 1565-1566.

Index

Contributors

J. S. Barlow, Department of Biological Sciences, Simon Fraser University, Burnaby, British Columbia, Canada

Gary J. Blomquist, Division of Biochemistry, University of Nevada, Reno, Nevada 89557

R. G. Bridges, A.R.C. Unit of Invertebrate Chemistry and Physiology, Department of Zoology, University of Cambridge, Cambridge, CB2 BEJ, England

H. M. Chu, Department of Entomology, University of Wisconsin, Madison, Wisconsin 53706

R. H. Dadd, Department of Entomology, University of California, Berkeley, California 94720

Yoshinori Fujimoto, Laboratory of Chemistry for Natural Products, Tokyo Institute of Technology, Nagatsuta, Midori-ku, Yokohama 227, Japan

Nobuo Ikekawa, Laboratory of Chemistry for Natural Products, Tokyo Institute of Technology, Nagatsuta, Midori-ku, Yokohama 227, Japan

Yoko Isaka, Laboratory of Chemistry for Natural Products, Tokyo Institute of Technology, Nagatsuta, Midori-ku, Yokohama 227, Japan

Spiro J. Louloudes, Plant Protection Institute, U.S. Department of Agriculture, Beltsville, Maryland 20705

J. E. McFarlane, Department of Entomology, Macdonald Campus of McGill University, Ste. Anne de Bellevue, Province of Quebec, Canada

T. E. Mittler, Department of Entomology, University of California, Berkeley, California 94720

Masuo Morisaki, Laboratory of Chemistry for Natural Prod-
cuts, Tokyo Institute of Technology, Nagatsuta, Midori-ku,
Yokohama 227, Japan

H. Noda, Shimane Agricultural Experiment Station, Izumo,
Shimane 693, Japan

D. M. Norris, Department of Entomology, University of Wisconsin, Madison, Wisconsin 53706

K.D.P. Rao, Department of Entomology, University of Wisconsin, Madison, Wisconsin 53706

Mertxe de Renobales, Division of Biochemistry, University
of Nevada, Reno, Nevada 89557

J. A. Svoboda, Insect Physiology Laboratory, Agricultural
Research Service, U.S. Department of Agriculture, Beltsville, Maryland 20705

Akihiro Takasu, Laboratory of Chemistry for Natural Products, Tokyo Institute of Technology, Nagatsuta, Midori-ku,
Yokohama 227, Japan

M. J. Thompson, Insect Physiology Laboratory, Agricultural
Research Service, U.S. Department of Agriculture, Beltsville, Maryland 20705

S. N. Thompson, Division of Biological Control, University
of California, Riverside, California 92521

Seppo Turunen, Department of Zoology, University of Helsinki,
00100 Helsinki 10, Finland

D. J. Van der Horst, Laboratory of Chemical Animal Physiology, State University of Utrecht, 3508 TB Utrecht, The
Netherlands

James L. Vaughn, Plant Protection Institute, U.S. Department
of Agriculture, Beltsville, Maryland 20705